Advances in Formal Design Methods for CAD

IFIP – The International Federation for Information Processing

IFIP was founded in 1960 under the auspices of UNESCO, following the First World Computer Congress held in Paris the previous year. An umbrella organization for societies working in information processing, IFIP's aim is two-fold: to support information processing within its member countries and to encourage technology transfer to developing nations. As its mission statement clearly states,

> IFIP's mission is to be the leading, truly international, apolitical organization which encourages and assists in the development, exploitation and application of information technology for the benefit of all people.

IFIP is a non-profitmaking organization, run almost solely by 2500 volunteers. It operates through a number of technical committees, which organize events and publications. IFIP's events range from an international congress to local seminars, but the most important are:

- the IFIP World Computer Congress, held every second year;
- open conferences;
- working conferences.

The flagship event is the IFIP World Computer Congress, at which both invited and contributed papers are presented. Contributed papers are rigorously refereed and the rejection rate is high.

As with the Congress, participation in the open conferences is open to all and papers may be invited or submitted. Again, submitted papers are stringently refereed.

The working conferences are structured differently. They are usually run by a working group and attendance is small and by invitation only. Their purpose is to create an atmosphere conducive to innovation and development. Refereeing is less rigorous and papers are subjected to extensive group discussion.

Publications arising from IFIP events vary. The papers presented at the IFIP World Computer Congress and at open conferences are published as conference proceedings, while the results of the working conferences are often published as collections of selected and edited papers.

Any national society whose primary activity is in information may apply to become a full member of IFIP, although full membership is restricted to one society per country. Full members are entitled to vote at the annual General Assembly, National societies preferring a less committed involvement may apply for associate or corresponding membership. Associate members enjoy the same benefits as full members, but without voting rights. Corresponding members are not represented in IFIP bodies. Affiliated membership is open to non-national societies, and individual and honorary membership schemes are also offered.

Advances in Formal Design Methods for CAD

Proceedings of the IFIP WG5.2 Workshop on Formal Design Methods for Computer-Aided Design, June 1995

Edited by

John S. Gero *and* **Fay Sudweeks** *(associate editor)*
Key Centre of Design Computing
University of Sydney
Sydney, Australia

SPRINGER-SCIENCE+BUSINESS MEDIA, B.V.

First edition 1996

Originally published by Chapman & Hall in 1996
Softcover reprint of the hardcover 1st edition 1996

ISBN 978-1-4757-4428-6 ISBN 978-0-387-34925-1 (eBook)
DOI 10.1007/978-0-387-34925-1

A catalogue record for this book is available from the British Library

Printed on permanent acid-free text paper, manufactured in accordance with ANSI/NISO Z39.48-1992 and ANSI/NISO Z39.48-1984 (Permanence of Paper).

CONTENTS

PREFACE

Designing is one of the most significant of human acts. It is one of the bases for change in our society. Designers are amongst the least recognised of society's change agents. Surprisingly, given that designing has been occurring for many millennia, our understanding of the processes of designing is remarkably limited. Part of our understanding of designing comes not only from studying human designers as they design but from postulating design methods which describe some aspect of the design process without claiming to model the processes used by human designers. The early approaches to design methods were prescriptive when applied to human designers. More recently, design methods have been formalised not as humano-centred processes but as processes capable of computer implementation. Amongst the goals of these endeavours are to develop a better understanding of the processes of designing, to develop methods which can be computerised and to aid human designers through the introduction of novel methods which have no human counterpart.

This move away from modeling human design processes and from prescriptive methods for human designers has opened up new areas in the development of formal design methods for computer-aided design. Although such methods have existed previously they were not fully explored because of their lack of humano-centricity. This nexus between a design method and its human use has been broken. This is not to imply that humano-centred design methods should not be developed or are of no significance, rather it is to suggest that they need to be augmented by these alternate approaches.

The working group of the International Federation for Information Processing (IFIP) which deals with computer-aided design (known as WG5.2) has been actively involved in a number of significant aspects of computer-aided design since the Group's inaugural conference in 1972. Over the last decade or so it has been pursuing artificial intelligence-based approaches to design methods and with issues concerned with geometric modeling. These approaches have spawned areas loosely called knowledge-based design, intelligent computer-aided design and knowledge intensive design. Formal design methods covers all these areas and more with its concern for the methodological process which is being modeled or developed and with the ability to characterise it in a formal sense.

The eleven contributions were presented under five topic headings and were followed by extensive discussions. A summary of the concepts traversed in the discussion of each session is provided after each group of contributions. The five sessions were:

1. evolutionary methods in design
2. generation and search methods in design
3. performance evaluation methods in design
4. formal support methods in design
5. design process methods

The overwhelming majority of the contributions describe design methods which have no direct human counterpart. Thus, these methods become adjuncts to human designers. This can be readily seen in those design methods which utilise large populations of designs rather than pursuing a single or very small number of nascent designs to fruition. These papers demonstrate the progress in formal design methods for computer-aided design.

All papers were sent to three reviewers for refereeing. A special thanks is due to these reviewers:

> Ömer Akin, Carnegie Mellon University, USA; Sylvie Boulanger, University of Sydney, Australia; Dave Brown, Worcester Polytechnic Institute, USA; Ken Brown, Carnegie Mellon University, USA; Jon Cagan, Carnegie Mellon University, USA; Jose Damski, University of Sydney, Australia; Chuck Eastman, University of California–Los Angeles, USA; Patrick Fitzhorn, Colorado State University, USA; Renate Fruchter, Stanford University, USA; Hans Grabowski, University of Karlsruhe, Germany; Mark Gross, University of Colorado, USA; Yehuda Kalay, University of California–Berkeley, USA; Vladimir Kazakov, University of Sydney, Australia; Larry Leifer, Stanford University, USA; Mihaly Lenart, University of Kassel, Germany; Mary Lou Maher, University of Sydney, Australia; Josiah Poon, University of Sydney, Australia; Michael Rosenman, University of Sydney, Australia; Tim Smithers, University of the Basque Country, Spain; George Stiny, University of California–Los Angeles, USA; Tapio Takala, Helsinki University of Technology, Finland; Hideaki Takeda, Nara Institute of Science and Technology, Japan; Toshiharu Taura, University of Tokyo, Japan; Jan Treur, Vrije Universiteit, The Netherlands; Enn Tyugu, Royal Institute of Technology, Sweden; Giorgio Valle, University of Milan, Italy

The undoubted success of this workshop was due to the support of the international program committee (listed at the rear of this volume) assisted by additional reviewers. The smooth management of the workshop was in large part due to the efforts of the local chair Carlos Zozaya-Goristza along with the conference manager Fay Sudweeks. Fay Sudweeks also acted as the associate editor of this volume, giving it its final shape and form in her inimitable manner. The assistance of the Key Centre of Design Computing, University of Sydney is acknowledged.

John S Gero
University of Sydney

AUTHOR ELECTRONIC ADDRESSES

Boulanger, S., sylvie@arch.su.edu.au
Brazier, F. M. T., frances@cs.vu.nl
Brown, K. N., kb58+@andrew.cmu.edu
Cagan, J., jcag+@andrew.cmu.edu
Gero, J. S., john@arch.su.edu.au
Grabowski, H., gr@rpk.mach.uni-karlsruhe.de
Joskowicz, L., josko@watson.ibm.com
Kalay, Y. E., kalay@ced.berkeley.edu
Kazakov, V. A., kaz@arch.su.edu.au
Lei, B., lei@race.u-tokyo.ac.jp
Lenart, M., michael@architektur.uni-kassel.de
Lossack, R.-S., lossack@rpk.mach.uni-karlsruhe.de
Maher, M. L., mary@arch.su.edu.au
Numata, J., numata@ssd.sony.co.jp
Poon, J., josiah@arch.su.edu.au
Rodgers, P. A., rodgerp@westminster.ac.uk
Rudolph, S., rudolph@isd.uni-stuttgart.de
Taura, T., taura@race.u-tokyo.ac.jp
Treur, J., treur@cs.vu.nl
van Langen, P. H. G., langen@cs.vu.nl
Weis, C., weis@rpk.mach.uni-karlsruhe.de
Zozaya-Gorostiza, C., zozaya@lamport.rhon.itam.mx

INTERNATIONAL PROGRAM COMMITTEE

LIST OF PARTICIPANTS

Brown, K., Design Computation Laboratory, Carnegie Mellon University, Pittsburgh PA 15213, USA

Cairó, O., División Académica de Computación, Instituto Tecnológico Autónomo de México, Rio Hondo #1, Col. Tizapán, San Angel, México DF 01000, México

Dong, A., Department of Mechanical Engineering, University of California–Berkeley, Berkeley CA 94720, USA

Gaffron, S., División Académica de Computación, Instituto Tecnológico Autónomo de México, Rio Hondo #1, Col. Tizapán, San Angel, México DF 01000, México

Gamboa, R., División Académica de Computación, Instituto Tecnológico Autónomo de México, Rio Hondo #1, Col. Tizapán, San Angel, México DF 01000, México

Gero, J., Key Centre of Design Computing, University of Sydney, NSW 2006 Australia

Gonzalez, M., División Académica de Computación, Instituto Tecnológico Autónomo de México, Rio Hondo #1, Col. Tizapán, San Angel, México DF 01000, México

Govela, A., División Académica de Computación, Instituto Tecnológico Autónomo de México, Rio Hondo #1, Col. Tizapán, San Angel, México DF 01000, México

Grabowski, H., Institute for Computer Application in Planning and Design (RPK), University of Karlsruhe, Kaiserstraβe 12, D-76128 Karlsruhe, Germany

Joskowicz, L., IBM Thomas J. Watson Research Center, PO Box 704, Yorktown Heights NY 10598, USA

Kalay, Y., Department of Architecture, University of California–Berkeley, Berkeley CA 94720, USA

Lei, B., Research into Artifacts, Center for Engineering, University of Tokyo, 4-6-1 Komaba, Meguro-ku, Tokyo 153, Japan

Lenart, M., Department of Architecture, University of Kassel, CAD-Zentrum, Diagonale 12, 34109 Kassel, Germany

Lossack, R.-S., Institute for Computer Application in Planning and Design (RPK), University of Karlsruhe, Kaiserstraβe 12, D-76128 Karlsruhe, Germany

Maher, M. L., Key Centre of Design Computing, University of Sydney NSW 2006, Australia

Rogers, P., SEMSE Design Group, University of Westminster, 115 New Cavendish Street, London W1M 8JS, United Kingdom

Rudolph, S., Institute of Statics and Dynamics, Stuttgart University, Pfaffenwaldring 27, D–70569, Stuttgart, Germany

Taura, T., Research into Artifacts, Center for Engineering, University of Tokyo, Komaba 4-6-1, Meguro-ku, Tokyo 153, Japan

van Langen, P., Department of Mathematics and Computer Science, Vrije Universiteit Amsterdam, De Boelelaan 1081a, 1081 HV Amsterdam, The Netherlands

Weis, C., Institute for Computer Application in Planning and Design (RPK), University of Karlsruhe, Kaiserstraβe 12, D-76128 Karlsruhe, Germany

Zozaya-Gorostiza, C., División Académica de Computación, Instituto Tecnológico Autónomo de México, Rio Hondo #1, Col. Tizapán, San Angel, México DF 01000, México

PART ONE

Evolutionary Methods in Design

1

FORMALISING DESIGN EXPLORATION AS CO-EVOLUTION

A Combined Gene Approach

MARY LOU MAHER, JOSIAH POON AND SYLVIE BOULANGER
University of Sydney, Australia

Abstract. This paper introduces a model for design exploration based on notions of evolution and demonstrates computational co-evolution using a modified genetic algorithm (GA). Evolution is extended to consider co-evolution where two systems evolve in response to each other. Co-evolution in design exploration supports the change, over time, of the design solution *and* the design requirements. The basic GA, which does not support our exploration model, evaluates individuals from a population of design solutions with an unchanged fitness function. This approach to evaluation implements search with a prefixed goal. Modifications to the basic GA are required to support exploration. Two approaches to implement a co-evolving GA are: a combined gene approach and a separate spaces approach. The combined gene approach includes the representation of the requirements and the solution within the genotype. The separate spaces approach models the requirements and the solutions as separately evolving interacting populations of genotypes. The combined gene approach is developed further in this paper and used to demonstrate design exploration in the domain of braced frame design for buildings. The issues related to the coding of the genotype, mapping to a phenotype, and evaluation of the phenotype are addressed. Preliminary results of co-evolution are presented that show how exploration differs from search.

1. Introduction

Most computer-based design tools assume designers work with a well-defined problem. The traditional treatment of design as two discrete phases: problem formulation and solution synthesis, is challenged by recent research. Though the view on discrete phases may be applicable to a simple and trivial design task, current research (Logan and Smithers, 1993; Corne, Smithers and Ross, 1994; Gero, 1994; Jonas, 1993; Navinchandran, 1991) has shown that design is an ill-structured problem and the discrete phases view is not a good (or correct) description of design.

Design is an iterative interplay to "fix" a problem from the problem space and to "search" plausible solutions from the corresponding solution space. The features and constraints in the current solution can become new criteria that lead to a redefined problem space, which in turn helps to generate a new design space. We call this phenomenon *exploration*. This is a phenomenon generally observed during conceptual design rather than detailed design. The present design tools from different domains, CAD or CASE, assume the designer has a clear picture of the problem and solution. These tools become a burden rather than help when the designer tries to alter the design. It is because these tools are not designed to cope with a change of requirements, change of design goals, change of assumptions, etc. Hence, if this phenomenon is not fully understood, it becomes difficult to know what kind of assistance could be provided to the designer during conceptual design.

The difference between search and exploration can be characterised by the input and output as illustrated in Figure 1. A typical search process generates a solution as its output with a well-defined problem as its input. However, an exploration process derives a problem and the corresponding solution from an ill-defined problem. It is not only because of the ill-defined nature of the problem that requires us to explore during design, the solution space also creates a need for this exploration.

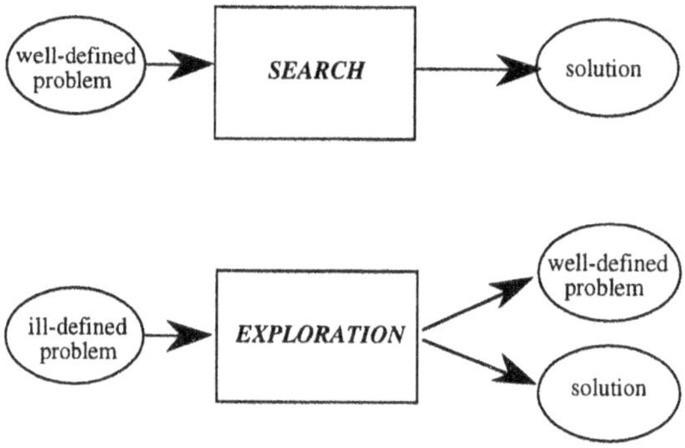

Figure 1. Input and output of search and exploration.

Genetic Algorithms (GAs) (Holland, 1962) provide an alternative to traditional search techniques by simulating mechanisms found in genetics. Three notions are borrowed from biological systems:

- the *phenotype*, which can be a living organism for biological systems or a design solution for design systems;
- the *genotype*, which is a way of representing or encoding the information which is used to produce the phenotype; and
- the *survival of the fittest*, which determines whether a genotype survives to reproduce.

In GA systems the genotype is usually represented as a binary string whose length varies with each application. For example, a genotype may look like: 001001101. GAs manipulate a representation (genotype) that differs from its expression (phenotype) in order to perform changes that couldn't be possible at the phenotype level. The genotype representation allows combination or mutation to occur in order to construct better strings. Some measure of fitness is applied to each string after combination and mutation to determine which strings participate in generating the next generation of the population.

A simple genetic algorithm considers a population of n strings and applies the operators: reproduction (or selection), crossover, and mutation in order to create the next generation. Reproduction is a process in which strings are copied according to their fitness function. Crossover is a process in which the newly reproduced strings are mated at random and each pair of strings partially exchanges information. Mutation is the occasional random alteration of the value of one of the bits in a string. Algorithms used to implement these processes are described in detail in Goldberg (1989).

In this paper a design process based on a genetic algorithm is presented which can model characteristics of explorative design: the search for problem definition as well as the search for solution. The use of an evolutionary system in which the genotypes represent alternative problem definitions and alternative solutions provides the basis for the co-evolution of problem space and solution space.

1.1. EXPLORATION IN DESIGN

Since design has been categorised as a problem solving activity (Simon, 1969), design is treated as a search of the solution space for a result. This idea has dominated the direction of artificial intelligence in design for some time. However, the validity of this hypothesis has been queried by recent work. For example, Corne, Smithers and Ross (1994) suggest that it is inappropriate to consider design as a search problem because a search problem requires a well defined problem space whereas a design problem is usually ill-structured. They propose design as "exploration" as follows:

> .. involves the construction and incremental extension of problem statements and associated solutions ..

Logan and Smithers (1993) further elaborate this definition that

> .. the formulation of the problem at any stage is not final ... As the design progresses, the designer learns more about possible problem and solution structures as new aspects of the situation become apparent and the inconsistencies inherent in the formulation of the problem are revealed. As a result, .. the problem and the solution are redefined...

Navinchandra (1991) defined exploration in the program CYCLOPS as

> .. Exploration is the process of generating and evaluating design alternatives that normally would not be considered..

He focuses on alternatives and this is achieved through criteria relaxation and criteria emergence. The relaxation is not constraints relaxation but a relaxation of the threshold value. This changing of threshold values causes a part of the solution space which is originally inside bound of the pareto curve to be explored. Solutions in this inside bound solution space can be examined as alternatives. Emergence described in CYCLOPS is a recognition activity. Criteria from precedent cases may be recognised to be relevant and interesting enough to apply to the current situation. The introduction of new criteria adds a new dimension for the designer to consider. The new criteria will be included to be part of the evaluation of design solution.

Another definition of "exploration" is provided by Gero (1994) that:

> .. Exploration in design can be characterised as a process which creates new design state spaces or modifies existing design state spaces...

This definition extends the "state space" concept of search (Simon, 1969), so that the state space is changed during exploration. This definition implies that the solutions in the given or predefined state space are insufficient for exploration. Gero continues to suggest that

> .. exploration precedes search and it, effectively, converts one formulation of the design problem into another .. Part of designing involves determining what to design for (function or teleology), determining how to measure satisfaction (behaviour), and determining what can be used in the final artefact (structure) ..

The definition relates exploration to search, indicating that exploration precedes search, and at the same time differentiates exploration from search.

Maher (1994) provides a definition which also relates search and exploration:

> .. search becomes exploration where the focus of the search changes as the process continues ..

This definition identifies search as a part of exploration, but not the same as exploration and also characterises the two as distinct, i.e. search has a definite goal while exploration doesn't. An approach to adaptive design, with the capability for exploration guided by the human designer, is described in Maher and Kundu (1994). Search as part of exploration cannot guarantee convergence because the design requirements change with the design solutions at the same time. However, convergence criteria could be externally defined and separate to the design requirements, recognising the fact that design usually completes when time has run out or factors external to the concerned problem.

For the remainder of this paper, we present a formal model of exploration. The model is illustrated in Figure 2 as the interaction of problem space and solution space. The problem space (or the functional requirements) is represented by P, and the solution space is represented by S. *Exploration* is defined as a phenomenon in design where P interacts and evolves with S over time.

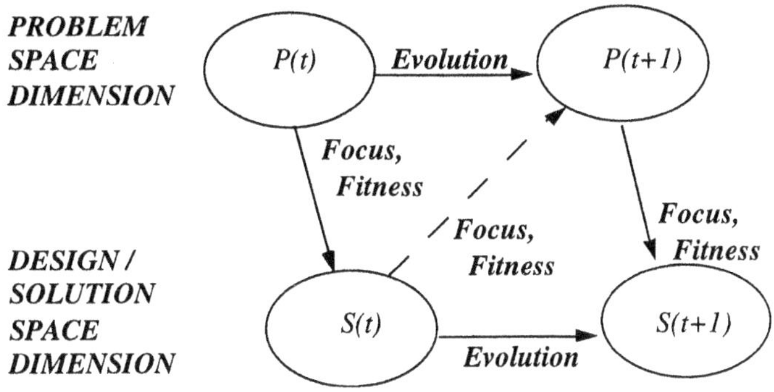

Figure 2. Problem-design exploration model.

The phenomenon of exploration as illustrated in Figure 2 has the following characteristics:

1. There are two distinct search spaces: Problem Space and Design Space.
2. These state spaces interact over a time spectrum.
3. Horizontal movement is an evolutionary process such that:
   (a) Problem space P(t) evolves to P(t+1), P(t+2), and so on;
   (b) Solution space S(t) evolves to S(t+1), S(t+2), and so on.
4. Diagonal movement is a search process where goals lead to solution. This can be *Problem leads to Solution* (downward arrow) or *Solution refocusses the Problem* (upward arrow).

The problem space P(t) is the design goal at time *t* and S(t) is the solution space which defines the current search space for design solutions. The solution space S(t) provides not only a state space where a design solution can be found, but it also prompts new requirements for P(t+1) which were not in the original problem space, P(t). This is represented by the dashed upward arrow from design space S(t) to problem space P(t+1). The upward arrow is an inverse operation where S(t) becomes the goal and a *search* is carried out in the problem space, P(t+1), for a *solution*. This iterative relationship between problem space and design space evolves over time.

This model of exploration depicts an evolutionary system, or in fact, two evolutionary systems. The evolutionary systems are the problem space and the solution space. The evolution of each space is guided by the most recent population in the other space. This model is called co-evolution and provides the basis for a computational model of design exploration. The basis for co-evolution is the simple genetic algorithm where special consideration is given to the representation and application of the fitness function so that the problem definition can change in response to the current solution space.

1.2. RELATED RESEARCH IN GENETIC ALGORITHMS

Genetic algorithms provide the basis for modelling evolutionary systems. The application of GAs to design include the solution to the truss design problem. The ten-member truss problem (Goldberg and Samtani, 1986) aims to find the optimal weight of each member for a given pre-determined configuration, such that the whole structure is stable and has a minimum weight. The configuration and the fitness function remains unchanged throughout the GA process. This represents a basic application of genetic algorithms to a design optimisation problem.

Watabe and Okino (1993) further study this problem by searching for structural shape as part of the problem. This is achieved by the introduction of new genetic operator called T-mutation. There are two types of T-mutation. The first one, T1-mutation, adds one new node to a randomly selected bar. The second one, T2-mutation, changes the topological structure without changing the number of nodes. The effect after the application of the T-mutation results in a new species which consists of individuals with different structural configurations. The changed configuration opens up a new solution space to search. This application of a genetic algorithm shows how the representation of the genotype determines the level at which the search occurs, in this case the search included a search for a configuration. However, the goal as defined by the fitness function remains to be the minimum weight configuration.

SAGA (Harvey, 1992), Species Adaptation Genetic Algorithms, allows the genotype to change in length as well as content so that species can emerge. He suggests that the notion of a search space is a metaphor when the question of "Where in this whole search space is the optimum?" is asked. However, this metaphor implies a space of pre-defined extent with a predefined goal. If a structure is to be evolved with potentially unrestricted capabilities, the simple GA, which has fixed length genotypes, is not an appropriate tool. The capability to represent a variable-length genotype is important to evolution. As the length increases, the population evolves as a species rather than global search. However, his model does not show how individuals from the solution space can affect the problem space.

Koza (1992) recognises the importance of co-evolution and suggests the term in biology is sometimes used to reflect the fact that all species are simultaneously co-evolving in a given physical environment. He uses Game Playing Strategy to elaborate on co-evolution, where two players in a game are represented as two populations of individuals. The fitness of a strategy of a player is measured by its performance against all strategies deployed by the other player. The fitness is, thus, a relative score. The performance of the two players continue to evolve with respect to the strategies by the opposing player. The mutual interactions and implicit relationships between players in a game are extended to a general conclusion as follows:

> In co-evolution, there are two (or more) populations of individuals. The environment for the first population consists of the second population. And conversely, the environment for the second population consists of the first population ... Co-evolution is a self-organising, mutually bootstrapping process that is driven only by relative fitness.

This provides a model for co-evolution where two solution spaces evolve in competition to each other, yet the goal remains the same. Co-evolution is affected by each search space defining the threshold for survival in the other search space. We present a co-evolutionary system in which the two spaces are not in competition with each other, yet they evolve in response to each other. Three differences between the co-evolution of Game Playing Strategy and the co-evolution of Problem-Design Space are:

1. The two populations in a game are opponents with the aim to beat each other, whereas in our co-evolution model, the aim is to explore the Problem Space and Design Space and to help each other to acquire better fitness values.
2. The purpose of co-evolution in the Game Playing Strategy is to measure how good an individual strategy can stand when played against various strategies by the opponent, while our co-evolution model aims to measure how good an individual from a population can satisfy (adapt) the expectations of individuals from another population.
3. The same fitness function is used for both spaces in the Game Playing Strategy, only the threshold for reproduction is changed in co-evolution, while our co-evolution model applies a potentially different fitness function to each space.

2. A Co-Evolutionary Process for Explorative Design

A simple GA, as shown below, is the basis for developing an evolutionary process model for explorative design.

When we apply the simple GA to the design process, we assume the process begins with an initial population of design genes that provide the information needed to generate a design solution. The evaluation determines which genotypes survive. The evaluation is performed by evaluating a fitness function and operates on the phenotype, which in the design process is the design solution. The processes of selection, crossover, mutation, and evaluation are the basis of the search for a design solution.

```
t = 0;
initialise genotypes in Population(t);
evaluate phenotypes in Population(t) for fitness;
while termination condition not satisfied do
```

```
t = t + 1;
select Population(t) from Population(t-1);
crossover genotypes in Population(t);
mutation of genotypes in Population(t);
evaluate phenotypes in Population(t);
```

Selection is a process in which individuals are copied according to their fitness function. This means that an individual with a higher value has a higher probability of contributing one or more offspring in the next generation. This operator is an artificial version of natural selection, a Darwinian survival of the fittest among individuals. In a natural population, fitness is determined by an individual's ability to survive. In the context of design, a fitness function representing the design requirements determines whether a design is suitable or not. Once an individual has been selected for reproduction, an exact replica of the individual is made. This individual is then entered into a mating place for further genetic operator action.

Crossover is a process in which the newly reproduced individuals are mated at random and each pair of individuals partially exchange information using a cross site chosen at random. For example if we consider the individuals A1 = 0110 | 1 and A2 = 1100 | 0 (where the separator symbol is indicated by |), the resulting crossover yields the following two new individuals A1' = 01100 and A2' = 11001. Crossover in a design process occurs when two design concepts are partially combined to form a new design concept.

Mutation is the occasional random alteration of the value of one of the bits in an individual. When used sparingly with reproduction and crossover, mutation is an insurance policy against loss of notions. In fact mutation plays a secondary role in the operation of GAs because the frequency of mutation to obtain good results in empirical GAs studies is on the order of one mutation per thousand bit transfers (Goldberg, 1989). Mutation has the potential to make small changes to a design concept, rather than a crossover process that makes large changes. We do not employ mutation in our co-evolutionary model of design.

Evaluation is a process of determining if a genotype continues in the next round of crossover. The termination condition is usually related to the evaluation, that is, when the evaluation of the population yields a suitable design, the process is terminated. Evaluation in the design process occurs by testing the performance of the design against relevant criteria. In the GA model of design, the fitness function is the basis for evaluation. The fitness function as a representation of design requirements can be predefined for the entire search process or it can be allowed to change as the genotype population changes. By changing the fitness function in response to the current population, the process models the ability of designers to change their focus when an interesting solution is found. This can be modelled as co-evolution of the design space and the performance space, where each space then becomes the population of genotypes for its own evolution and the fitness function for the other space.

The co-evolution of the design genes (solution space) and the fitness function (problem space) provides a model for design as exploration. Two approaches to representing coevolution are:

- **CoGA1**: A single composite genotype is formed by the combination of a problem requirements and a design solution. The fitness function is defined locally for each design solution.
- **CoGA2**: The two spaces are modelled as two sets of genotypes and phenotypes: one for modelling problem requirements and one for modelling design solutions. The current population of each space provides the fitness measurements for the other.

2.1. COGA1: COMBINED GENE APPROACH

This first co-evolving algorithm has two modifications to the basic GA:

1. The fitness function (problem part, P) and design solution (solution part, S) are put into one genotype.
2. There are two phases of crossover-evaluation operations in each generation instead of the convention of one phase.

The algorithm, CoGA1, is shown below.

```
CoGA1
t = 0;
initialise genotypes in Population(t);
evaluate phenotypes in Population(t) for fitness;
while termination condition not satisfied do
        t = t + 1;
        select Population(t) from Population(t-1);
        /* Phase 1: from S to P */
                crossover genotypes in Population(t) at Performance_space;
                evaluate phenotypes in Population(t);
        /* Phase 2: from P to S */
                crossover genotypes in Population(t) at Design_space;
                evaluate phenotypes in Population(t);
```

Inside the repeating loop, there are two phases of GA operations for each generation. If no satisfactory solution is found in previous operations with the stated problem, the problem is revised to give new dimensions for the solution space. Hence, the first phase corresponds to the shift of attention of fitness function when a solution space is given, i.e. the upward arrow from *S* to *P* in our model of problem-design exploration. In phase 1, crossover occurs in the problem part of the genotype, as illustrated in Figure 3(a). For example, the crossover point to the parent genotypes cut the problem part to P11 and P12, and P21 and P22. For the same solution carried forward from the previous phase, the fitness is evaluated using a different fitness function, i.e. the same S1, which is evaluated by P11 and P12 in parent genotype, is evaluated by P11 and P22 in the new recombined child genotype. The fitness value for each design solution represents a local fitness.

After the problem is revised, the second phase relates to the search for a solution with the reformulated fitness function from Phase 1. This corresponds to the downward arrow from *P* to *S* in the model, as shown in Figure 2. Crossover occurs on the design solution part of genotypes: the S11, S12, S21 and S22 in Figure 3(b). The fitness of a design solution is not evaluated by a common global fitness function, but by the fitness function defined as the problem part in the same genotype. In other words, the fitness score of each genotype is again a local fitness value. In our example, the offspring which has solution part composed of S11 and S22 is evaluated by P1; while the other offspring, which has S21 and S12 in its solution part, is evaluated by P2.

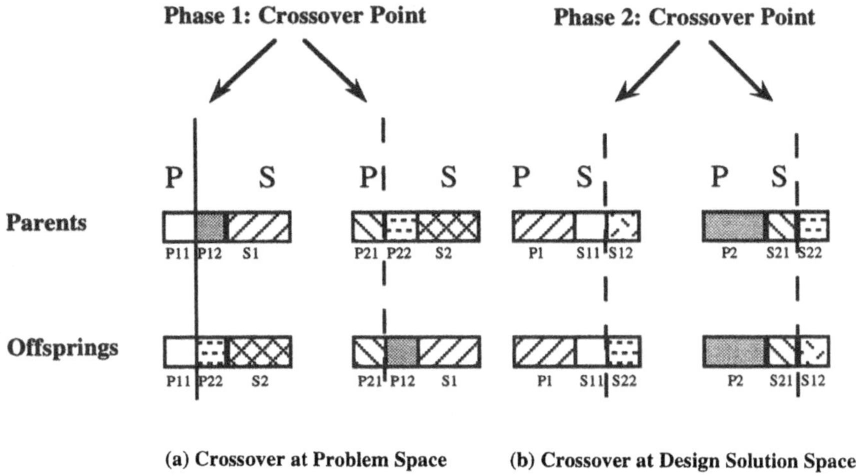

Figure 3. Crossover operation for CoGA1.

After phase 2 the solutions are used to check the termination condition to determine whether another generation is necessary. Currently, the termination condition is defined as a fixed number of generations, meaning that the exploration stops when a predetermined amount of time has passed. However, the termination condition can be any globally defined condition that does not evolve in response to the alternatives found in the solution space.

2.2. COGA2: TWO INTERACTING POPULATIONS

A second approach to co-evolution is to maintain separate spaces of genotypes for problem requirements and design solutions. As illustrated in Figure 4,

CoGA2 uses the current selection from each population to be the fitness function for evaluating the individuals in the other population. The problem requirements is modelled as a collection of criterion, where each criterion is represented as a genotype in the Problem Space.

Every problem criterion genotype has a label and a weighting (i.e. the genotype has a length of 2). A problem is, thus, a combination of individual genotypes with their current weights. If we allow the crossover operator to cut and paste a different weight to a criterion, followed by selecting a random number of genotypes, these problem criteria will collectively define a problem which has a different perspective and emphasis to be solved. The fitness of a solution is defined by the current collection of criterion. In the other direction, the fitness of a criterion is defined by the number of times that criterion is satisfied in the current collection of individuals in the Solution Space.

The CoGA2 starts with initialising the two populations which represent problem and solution. An initial evaluation of individuals from the Solution Space is performed using the initial design requirements as defined by the user. The initial evaluation of the Problem Space is performed based on the selected individuals from the Solution Space. The termination condition is checked and the pattern of "phases" appear in CoGA2 as well. Each phase in CoGA2 corresponds to a different evaluation function, rather than to a different crossover operation as in CoGA1. The CoGA2 algorithm is shown below.

CoGA2:

```
t = 0;
initialise genotypes in Problem_space(t);
initialise genotypes in Solution_space(t);
initial-evaluate phenotypes in Solution_space(t) for fitness according
        to user's initial requirements;
initial-evaluate phenotypes in Problem_space(t) for fitness according
        to user's initial requirements;
while termination condition not satisfied do
        t = t + 1;
        /* Phase 1: from S to P */
                select Problem_space(t) from Problem_space(t-1);
                crossover genotypes in Problem_space(t);
                evaluate phenotypes in Problem_space(t) for fitness
                        according to selected individuals from
                        Solution_space(t-1);
        /* Phase 2: from P to S */
                select Solution_space(t) from Solution_space(t-1);
                crossover genotypes in Solution_space(t);
                evaluate phenotypes in Solution_space(t) for fitness
                        according to selected sample of individuals from
                        Problem_space(t-1);
```

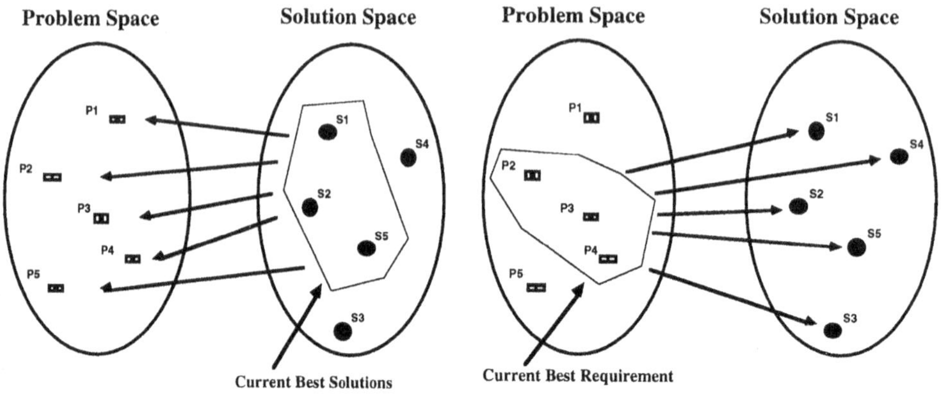

(a) Fitness evaluation of problem space (b) Fitness evaluation of solution space

Figure 4. Fitness evaluations in CoGA2.

3. The Combined Gene Approach for Braced Frame Design

The design domain selected to demonstrate the combined gene approach to co-evolution is steel braced frames. The domain provides a range of design solutions that can be described geometrically and more than one design focus. The design of braced frames for buildings is done by a structural engineer with some constraints/requirements imposed by the architect's layout of the building. Two alternatives for the design focus considered here are: design for architectural compatibility and design for structural resistance. These alternatives are represented as more specific design performance criteria, as listed below.

1. Architectural compatibility
   - integration of frame into bay layout
2. Structural resistance
   - efficiency of the frame
   - integrity of the frame

Architectural compatibility is evaluated according to how well the braced frame fits the bay layout as predefined by an architect in building design. Structural efficiency is evaluated in one of two ways: as a measure of the efficiency of the use of material in the braced frame or as a measure of the integrity or ductility of the braced frame. The result of this formulation of the design problem is that there are three possible ways to evaluate the performance of a braced frame: how well the frame fits the bay layout; the structural efficiency of the frame; and the integrity of the frame. The implication of this formulation is that one focus may lead to a different set of designs to the set that would result from another focus.

In addition to the design focus, the design of braced frames is defined by the geometric requirements imposed by the architectural design of the building. The problem definition is formalised through the values of a set of parameters, as shown below.

parameter	*range*
number of storeys (n_storey)	5-20 storeys
storey height (h_storey)	3.0 - 4.5 m
bay width (w_bay)	5.0 - 20.0 m
design focus (criteria)	compatibility, efficiency, or integrity

The braced frame solution description is represented by seven features, where each feature may take on one of a range of values :

feature *range*	
type of panel (type)	cross, single diagonal, chevron, diamond
flip within panel (flip)	none, horizontal, vertical
mirror above panel (mirror)	no, yes
eccentricity in panel	
—at top beam level (eccent_1)	none, 1/8 or 1/4 width of frame
—at bottom beam level (eccent_2)	none, 1/8 or 1/4 width of frame
multi-storey panel (m_storeys)	1, 2 or 3
ratio of bay size to panel width(c_panel)	0.5 to 2.0

A solution is generated from the above features by taking the type of the panel (i.e. cross, single diagonal, chevron, or diamond) and applying the transformations defined by the values of the remaining features. Figure 5 illustrates these transformations on a panel. Representing the alternative design solutions as a set of transformation features allows for a very rich generation of alternatives by concentrating on the geometric manipulation of the panel. These features provide the basis for the genotype representation.

Additional parameters, not coded directly in the genotype, are derived from the above features and contribute to the description and the evaluation of the frame. These parameters are shown below. The combination of the features in the genotype and the derived parameters below provide the basis for a phenotype description. This distinction is important because the genotype is the basis for producing the next generation, the phenotype is the basis for evaluation and the survival of the fittest.

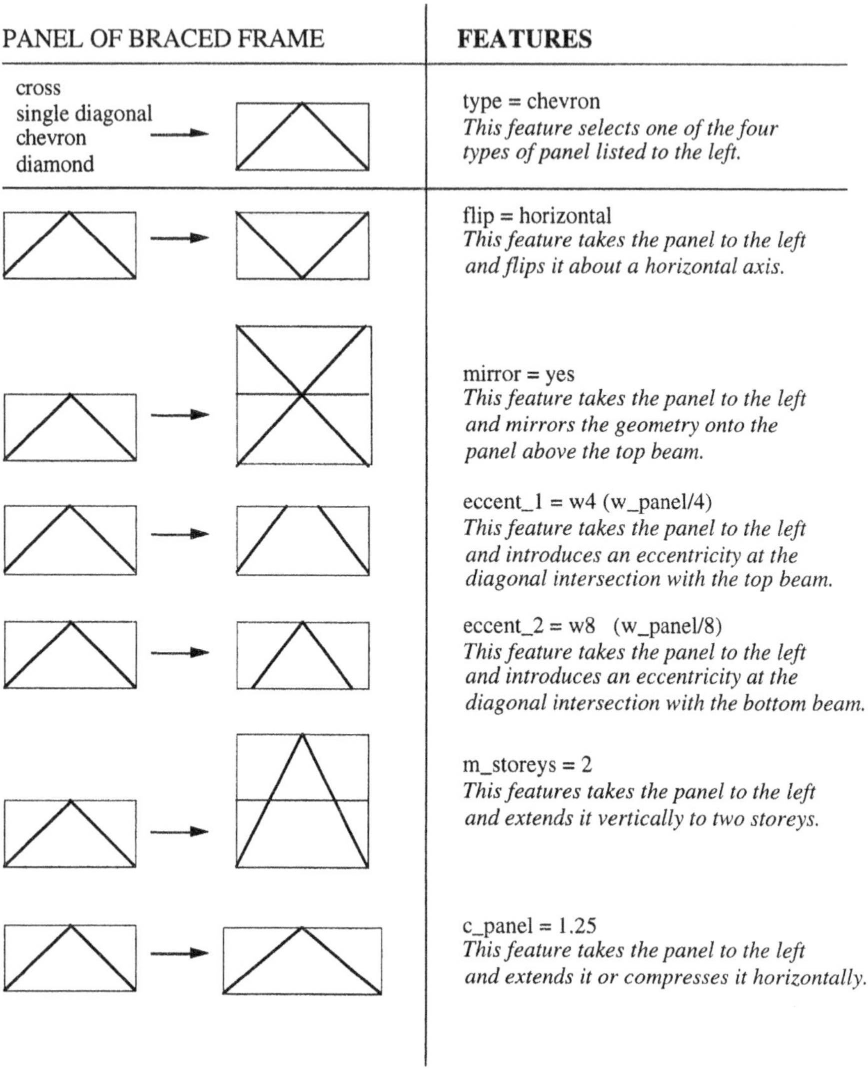

Figure 5. Example of braced frame features and their use in defining an alternative.

parameter	*equation*
width of panel	w_panel = c_panel x w_bay
width of frame	W_frame = w_panel
height of panel	h_panel = m_storeys x h_storey
height of frame	H_frame = n_storey x h_storey
aspect ratio of panel	h_panel / w_panel
aspect ratio of frame	H_frame / W_frame
zone of feasibility	z_feasib
coefficient of ductility	c_ductility

There is an important distinction between three entities: the frame, the bay and the panel. A frame is an assembly of adjacent panels covering the full height and full or partial width of the building. A bay relates to the layout of beams in the building within which the frame is integrated and represents the distance between two columns (vertical elements). A panel is a unit of the frame which may be repeated vertically and/or horizontally.

In this example we define four possible fitness functions: one evaluates the difference between the initial problem parameters and current parameters; the other three evaluate the fitness of the current features for a design focus. The first evaluation function is defined in order to follow the variation of the population from the initial requirements, remembering that during exploration the design requirements may change. The other evaluation functions are defined to measure the performance of the phenotype according to a selected design focus, remembering that the combined gene approach to co-evolution includes the definition of performance criteria within the genotype.

The evaluation functions return a value between 0 and 1. The functions themselves are based on heuristic evaluations of the alternatives, that is, no mathematical modelling or analysis of the alternatives is performed. When a function returns a value of 0.0, the alternative will not survive to reproduce in the next generation. The lower the value, the less probability of its use for the next generation. The details of the evaluation functions for braced frame design are given in the Appendix.

3.1 REPRESENTATION OF BRACED FRAME AS GENOTYPE AND PHENOTYPE

Figure 6 illustrates a drawing of a frame and its associated formal description for this demonstration. Each frame is considered in the context in which it was designed; this is a consequence of the combined gene approach where the genotype includes both the problem description and the design solution. Consider a designer who needs to find a suitable frame for an 8-storey building with 3.7 m storey height using 12 m bays, and the designer chooses to focus on the structural efficiency of the frame. By letting the problem parameters and the solution features vary, it is possible to explore a wide range of potential solutions. One suitable solution, as shown in Figure 6, is a cross-braced panel, over two storeys, which is 7.8 m wide. It is interesting to note that the initial bay width was 12 m. This solution satisfies structural efficiency very well with a value of 0.95.

The genotype of a braced frame comprises the initial problem parameters, including the design focus, and the features of the design solution. Each genotype is represented as a fixed length binary string. The bits in a genotype are grouped into chunks, where each chunk is a contiguous block of bits. The position of each chunk uniquely identifies a parameter or feature of the frame and the binary string in the chunk maps onto the value of that feature/parameter.

In this example, each chunk occupies 5 bits, i.e. each chunk represents 32 possible values.

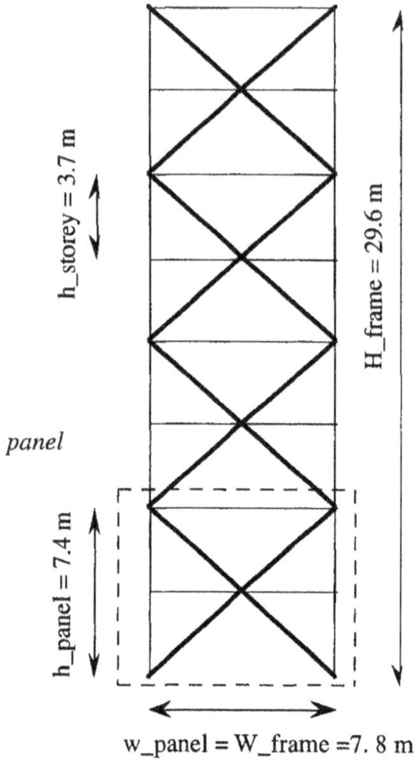

Initial problem parameters
criteria = structural efficiency
n_storey = 8
h_storey = 3.7 m
w_bay = 12 m

Final solution features
type = cross
flip = vertical
mirror = yes
eccent_1 = none
eccent_2 = none
m_storeys = 2
c_panel = 0.65

Derived parameters
ar_panel = h_panel/w_panel = 0.95
ar_frame = H_frame/W_frame = 3.8

Evaluation
structural efficiency = 0.95

Figure 6. Example of a complete braced frame description.

In our combined gene approach, a genotype consists of two parts: the problem part and the solution part. The values found in the chunks among the problem part represent the design requirements and focus while the solution part provides values to describe the geometry of the solution. The template of a genotype is shown in Figure 7, the positions before the double vertical line are for problem parameters, and those which are found after the lines stand for solution features.

criteria	n_storey	h_storey	w_bay	type	flip	mirror	eccent_1	eccent_2	m_storeys	c_panel

Figure 7. Genotype template of a braced frame.

One genotype in a population is represented as a binary string as shown in Figure 8. The figure has three parts: the top part shows the genotype as a binary string, the second part shows the mapping from binary string to attribute-value pairs and the third part shows the distribution of values in a 5 bit chunk,

Since the features/parameters do not have 32 possible values, a mapping is defined to identify the value to be used. In mapping some features, we used a biased coding scheme such that certain values of an attribute are more preferred than the other options. For example, there are three possible values for the feature "eccentricity in panel": none, w/8, and w/4. The value none is preferred over the introduction of an eccentricity. In the mapping we have given the value none a probability of 0.5, and w/8 and w/4 probabilities of 0.25. The biased mapping for "eccentricity in panel" is shown in the "chunk" in Figure 8, where the first 16 values for the parameter are mapped onto none, etc.

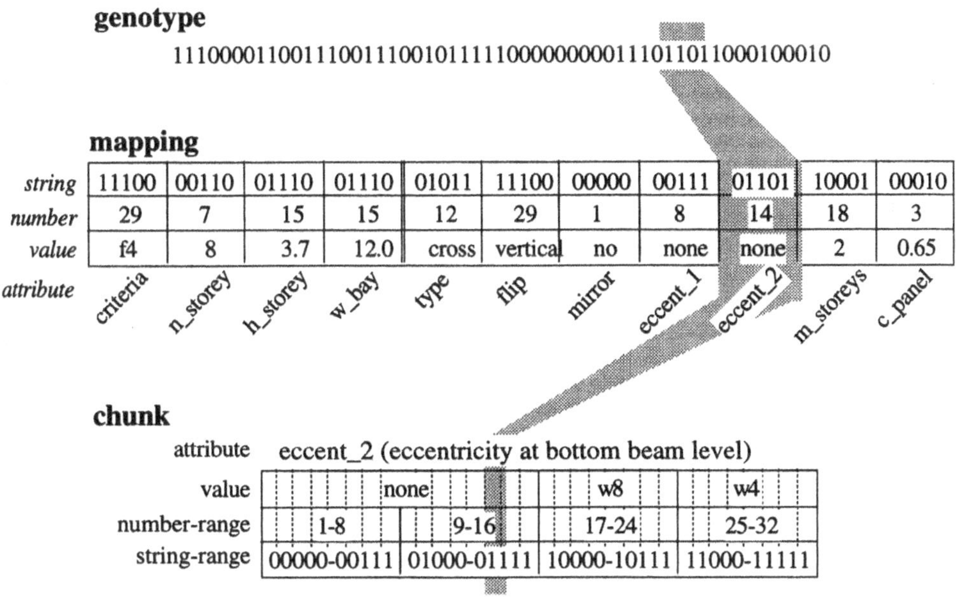

Figure 8. A genotype representation of a braced frame.

The phenotype is the representation of the braced frame that is evaluated. The phenotype representation is defined through a mapping from the genotype representation followed by a derivation of additional parameters. Basically, the phenotype is represented as a set of attribute-value pairs. The mapping process involves three steps:

1. Divide the genotype into chunks.
2. Convert the binary string in each chunk to a decimal value.
3. Map this value to the appropriate symbolic value.

The first two steps are completed by a fixed mapping procedure, while the last step is different for each attribute. The mapping takes into consideration the biased coding of the feature. Hence, the feature "mirror" has two options, no or yes, and there is no bias. Therefore, the chunk number of the mirror feature is determined using the following rules:

if chunk_number < 16
then mirror = no
if chunk_number >= 16
then mirror = yes

The mapping is defined in such a way as to be consistent in either direction. Once the mapping is completed, values for additional parameters can be derived from these mapped values. For example, the aspect ratio of the panel (ar_ panel) is defined to be h_panel/w_panel. The value of h_panel is defined to be m_storeys x h_storey, etc.

The phenotype is given a fitness value according to one of the four evaluation functions introduced above. The evaluation functions are:

f0: distance from initial requirements
f2: compatibility with bay layout
f3: structural efficiency
f4: structural integrity

The details of how the braced frame phenotype is assigned a value for each of these functions is given in the Appendix. These evaluation functions provide the value used to determine the probability of the genotype being used in the next generation. The initial population is evaluated with a special evaluation function in order to start with a population that is relevant to the initial requirements. This function ensures that the designer's focus is considered before the exploration process begins. The initial fitness function is defined below, where finit is the value of the evaluation function that matches the designer's original focus.

Fi: initial fitness function = f0 + finit

where finit = f2, f3 or f4

In addition to these evaluation functions, another evaluation function has been defined to follow the performance of the population. A global fitness function, Fg, is defined to be the average of the values of the four evaluation functions. This value is not used in the selection process since CoGA1 is guided by a local fitness function.

3.2. RESULTS OF CO-EVOLUTION

This section presents the results of using CoGA1 on the braced frame problem formulation. We decided to generate the initial population of genotypes randomly and to run the co-evolution with a genotype population of 100. The algorithm uses roulette selection, i.e. fitness proportionate selection. Each new population is generated with a probability of crossover being 1.0. No mutation is applied at this stage.

In run A, the design focus is defined to be structural integrity for a building that is 20 storeys high, each storey has a nominal height of 3.8 m, and the suggested bay width is 12 m. In summary, the design requirements are:

Criteria (i_criteria)	= structural integrity
No. of storey (i_n_storey)	= 20
Height of storey (i_h_storey)	= 3.8 m
Bay width (i_w_bay)	= 12 m

The best solution proposed after 23 generations is a very reasonable design solution (Figure 9).

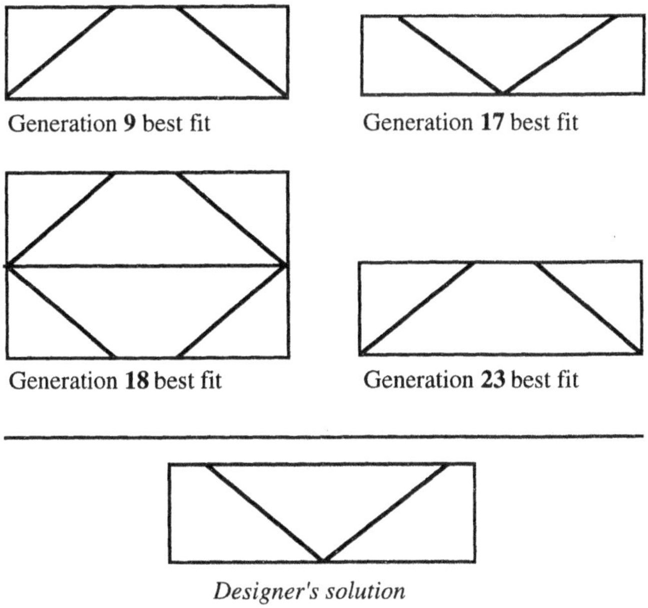

Figure 9. Intermediate and final solutions for run A.

The chevron with an eccentricity of w/4 at the top beam level satisfies the initial focus (integrity) very well and does not deviate significantly from the initial problem parameters. Although the final best fit is different than the designer's solution, it remains compatible with the concept of an eccentrically braced frame chevron. Additionally, an interesting best fit emerges at generation 18, with an eccentricity introduced in what appears to be a diamond. In reality, this solution resulted from both flip and mirror features acting on a single chevron panel. Finally, the best fit of generation 23 was first proposed at generation 9.

This indicates that convergence was not conclusive at generation 9 and more exploration was required. The pursuit of a moving target (or change in focus) resulted in a variety of alternatives before the final best fit was reached. Hence, the evolution of alternatives is also reflective of the exploratory approach expressed by the interplay between problem and solution spaces.

The algorithm converged after 23 generations with a design focus of structural integrity. The chart shown in Figure 10 is a summary of the percentage of the population that had each evaluation function as its local fitness (design focus) across all 23 generations until convergence. The chart shows the variation in focus for the braced frame solutions. This demonstrates that our goal of exploration as a change in focus during the design process occurred through the rise and fall of the proportion of the population that used each design focus.

The graph in Figure 11 shows the overall performance of each generation as defined by the value of Fg. Fg did not influence the fitness of the phenotypes, it is only shown here to measure overall performance in a computational method that allows design exploration. It is interesting to note the fluctuation of the value of the best fit solution vs the average value of the entire population. This implies that the best fit influences the fitness of the entire population, regardless of its local fitness function. In the graph below the population at t=0 represents the fitness of the randomly generated genotypes, the population at t=1 and greater represent phenotypes that have been evaluated for selection. Note that the global fitness after t=1 increases almost monotonically.

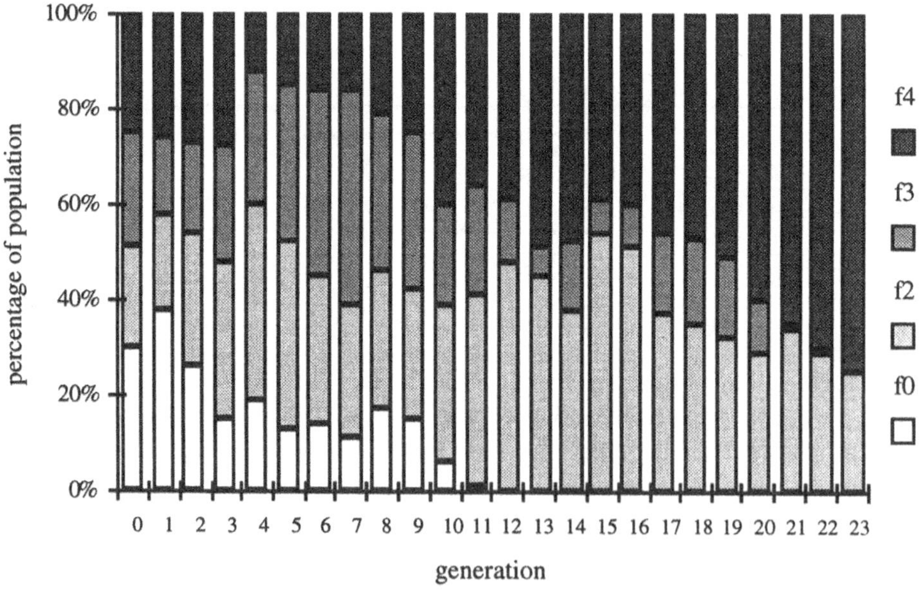

Figure 10. Evolution of design focus for run A.

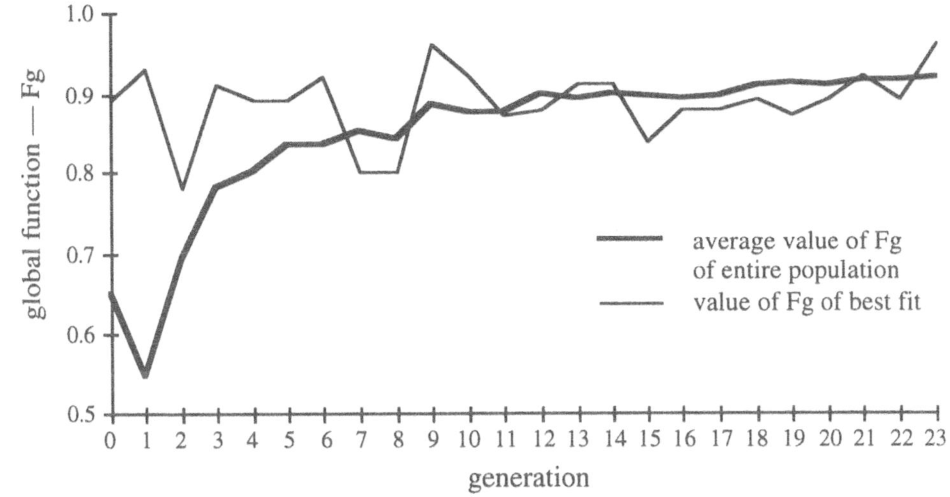

Figure 11. Overall performance of population during co-evolution for run A.

The chart shown in Figure 12 is a summary of the results of a different set of initial requirements, run B. It is shown here to illustrate a different evolution of the design focus through a different distribution of phenotypes with a specific design focus in each generation. This distribution is remarkably different to Figure 10. In Figure 12 one design focus dominates at convergence, where in Figure 10 two different foci are present at convergence. In fact, as shown in Figure 12, design focus f3 starts to dominate at generation 18 and at convergence the dominate design focus is f4. This further demonstrates design exploration through a change in design focus.

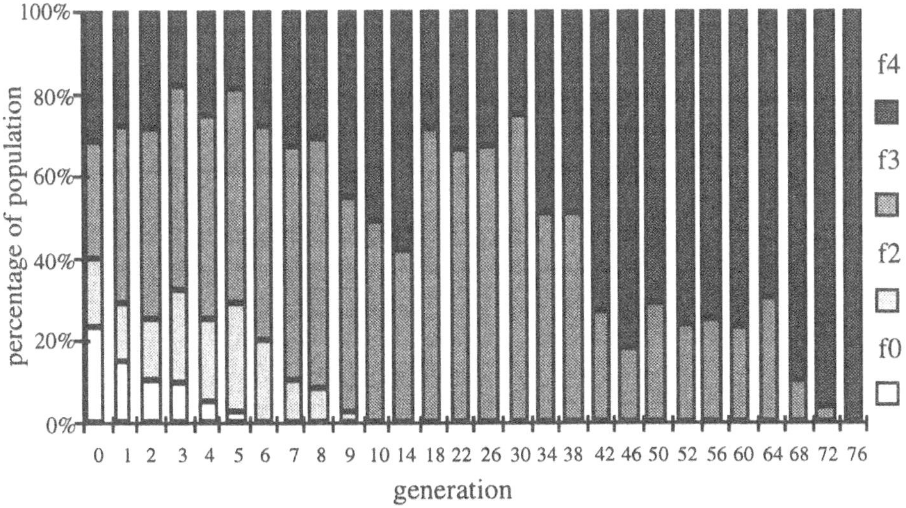

Figure 12. Evolution of design focus for run B.

4. Discussion

The development and implementation of a model of exploration as co-evolving spaces has enabled a computational approach to changes in design focus. The implementation as a modified genetic algorithm raises a number of issues.

The quality of the initial population has important effects on how the evolution progresses. If generation 0 is not filtered by Fi, the evolution converges quickly to the focus functions held by strong individuals of the initial population. This suggests that the random number selected for a particular run can be very influential on which functions dominate at convergence. If generation 0 is filtered by Fi, this tendency is present but not as strong. For instance, applying the initial Fi may reduce the overall Fg value from generation 0 to generation 1, but at least, finit has a better probability of being favoured although its value at generation 1 is low. This may influence the number of generations until convergence but not the pattern of exploration.

The quality of the evaluation functions in terms of how they relate to each other and not just how adequately they attribute a value within one evaluation function is important. A heterogenous quality of functions may have an immediate influence on which focus function takes over for convergence. In one case, applying a weight < 1.0 to f4 made the next run shift from f4 to f3 as a final focus. Although this is an important parameter to desensitise, at the same time, a shift does not necessarily lead to low values for other functions.

The relative ease of satisfying a particular design focus changes the way exploration works in this approach. In the first several runs we included an additional design focus for architectural compatibility which checks whether the frame accommodates openings as specified by the architect. We called this fitness function f1. For the current formulation it was much easier to satisfy f1 than the other fitness functions. As a result, all runs were similar - f1 dominated and the best fit solution was always one that satisfied f1. We removed f1 for consideration until we could better calibrate its evaluation to be similar to the other functions for design focus. This implies that the coding of the design focus functions need to be considered in terms of their ease of satisfaction relative to the other functions.

Pool size does not appear to have marked influences on the final results. Rather, a higher number of individuals in the population seems to reduce the number of iterations required for convergence. However, similar patterns of exploration emerge.

The mapping approach used has a non-conclusive effect on the final results. Scaling possible values of one feature from 7 values to 32 values, while retaining the same minimum and maximum values, has caused an important shift favouring the evaluation function which uses that modified feature. However, no clear pattern has surfaced yet.

The parameters which need to be analysed more carefully are: the sensitivity of threshold values used on Fi to influence the initial population; the absence or presence of f0 as a problem evaluation function which co-exists with solution evaluation functions such as f2, f3 and f4 ; the interdependence of evaluation functions in terms of their maximum and obtainable values; the ease with which designs can attain high evaluation values; the random number applied to a run; the bias and granularity applied to map a parameter's symbolic values unto a scale of 32.

5. Conclusion

The co-evolution of problem space and solution space shows promising results on how changes in focus provide a framework for explorative design. In this paper we present a model of exploration, a method of exploration that uses co-evolution, and an implementation of this method for the design of braced frames. Our preliminary results show that design exploration can be modelled as computational methods. Our preliminary results also show that the representation of the design domain can result in an unstable method - that is, one that produces a different result for the same problem in two different runs. This instability may be a desirable feature since it provides a computational model that is as unpredictable as human designers. However, we need to test the model further to understand why there is instability and how the genotype, phenotype, and evaluation functions influence the nature of the exploration.

In addition to further testing CoGA1, the model of exploration needs to be taken further to develop CoGA2, the separate spaces formulation, and to consider formulations that allow a design focus to emerge that has not been identified before the generation of alternatives.

Acknowledgments

This work is supported by the Australian Research Council and an Australian Postgraduate Research Award.

References

Chien, E.: 1988, *Multi-Storey Steel Building Design Aid*, Canadian Institute of Steel Construction, Toronto, Canada.

Corne, D., Smithers, T. and Ross, P.: 1994, Solving design problems by computational exploration, *in* J. S. Gero and E. Tyugu (eds), *Formal Design Methods for Computer-Aided Design*, North-Holland, Amsterdam, pp. 293-313.

Gero, J. S.: 1994, Towards a model of exploration in computer-aided design, *in* J. S. Gero and E. Tyugu (eds), *Formal Design Methods for Computer-Aided Design*, North-Holland, Amsterdam, pp. 315-336.

Gero, J. S., Louis, S. J. and Kundu, S.: 1994, Evolutionary learning of novel grammars for design improvement, *Artificial Intelligence for Engineering Design, Analysis and Manufacturing*, **8**, 83–94.

Goldberg, D. E.: 1989, *Genetic Algorithms: In Search, Optimization and Machine Learning*, Addison-Wesley, Reading, MA.

Goldberg, D. E. and Samtani, M. P.: 1986, Engineering optimization via genetic algorithm, *Proceedings of the Ninth Conference on Electronic Computation*, pp. 471–482.

Harvey, I.: 1992, Species adaptation genetic algorithms: A basis for a continuing SAGA, *in* F. J. Varela and P. Bourgine (eds), *Toward a Practice of Autonomous Systems, Proceedings of First European Conference on Artificial Life*, MIT Press/Bradford Books, Cambridge, Mass.

Holland, J. H.: 1962, Concerning efficient adaptative systems, *in* M. C. Yovits, G. T. Jacobi, and G. D. Goldstein (eds), *Self-organizing Systems*, Spartan Books, pp. 215-230.

Jonas, W.: 1993, Design as problem-solving? or: Here is the solution - What was the problem? *Design Studies*, **14**(2), 157-170.

Koza, J. R.: 1992, *Genetic Programming: On the Programming of Computers by Means of Natural Selection,* MIT Press, Cambridge, Mass.

Logan, B. and Smithers, T.: 1993, Creativity and design as exploration, *in* J. S. Gero and M. L. Maher (eds), *Modelling Creativity and Knowledge-Based Creative Design*, Lawrence Erlbaum Associates, Hillsdale, NJ, pp. 139–175.

Maher, M. L.: 1994, Creative design using a genetic algorithm, *Computing in Civil Engineering*, ASCE, pp. 2014-2021.

Maher, M. L. and Kundu, S.: 1994, Adaptive design using a genetic algorithm, *in* J. S. Gero and E. Tyugu (eds), *Formal Design Methods for Computer-Aided Design*, North-Holland, Amsterdam, pp. 245-263.

Michalewics, Z.: 1992, *Genetic Algorithms + Data Structures = Evolution Programs*, Springer-Verlag, Berlin.

Navinchandra, D.: 1991, *Exploration and Innovation in Design*, Springer-Verlag, New York.

Simon, H. A.: 1969, *The Sciences of the Artificial*, MIT Press., Cambridge, Mass

Watabe, H. and Okino, N.: 1993, Structural shape optimization by multi-species genetic algorithm, *in* C. Rowles, H. Liu and N. Foo (eds), *Proceedings of the 6th Australian Joint Conference on Artificial Intelligence (AI'93)*, Melbourne, Australia, pp. 109–116.

Appendix: Evaluation functions for braced frame design

There are four evaluation functions used in this example for evaluating the performance of a braced frame:

f0: measure of closeness to initial requirements,
f2: measure of conformance to bay layout
f3: measure of structural efficiency
f4: measure of structural integrity

The first function, f0, uses a combination of three functions, each concentrating on measuring the distance between the current phenotype and the one problem parameter in the initial requirements. The value of f0 is the average value of these three functions: f01, for the parameter n_storey, f02, for the parameter h_storey, and f03, for the parameter w_bay. The value of each function is determined using a normal distribution shape applied to n_storey, h_storey and w_bay, as shown in Figure A.1. The vertical axis shows the value of f0 and the horizontal axis shows the value d_parameter which corresponds to the normalised difference between the phenotype and the parameter. For each parameter, an acceptable deviation was assumed. The calculation of the value of f0 is shown below.

```
/* acceptable deviations from the initial problem parameters */
bldg->sig_n_storey = 2      /* storeys */
bldg->sig_h_storey = 0.2    /* metres  */
bldg->sig_w_bay = 1.5       /* metres  */

/* normal distribution shape function used to assess the difference */
bldg->f01 = exp(-pow((bldg->n_storey - bldg->i_n_storey),2)/
            (2 * pow(bldg->sig_n_storey,2)))
bldg->f02 = exp(-pow((bldg->h_storey - bldg->i_h_storey),2)/
            (2 * pow(bldg->sig_h_storey,2)))
bldg->f03 = exp(-pow((bldg->w_bay - bldg->i_w_bay),2)/
            (2 * pow(bldg->sig_w_bay,2)))

/* sum of the squares method for obtaining function f0 */
bldg->sum_square = pow(bldg->f01, 2) + pow(bldg->f02, 2) + pow(bldg->f03, 2)
bldg->f0 = pow(bldg->sum_square, 0.5)
```

The evaluation function, f2, is concerned with evaluating how well the phenotype satisfies the design focus of architectural compatibility. This function measures how well the panel configuration merges with the bay size as defined by the architect. Designers are often faced with fitting a frame into a proposed layout. In order not to restrict the geometry immediately to the proposed bay width, some variation is permitted. This variation is represented by the panel coefficient which ranges from 0.5 to 2.0. The panel width is obtained by multiplying the bay width by the panel coefficient. This can simulate the interaction between owner/architect/engineer during the preliminary phases. An architect proposes a layout, the engineer proposes a frame, the emphasis on architectural compatibility and structural efficiency might be discussed, resulting in a layout change, or a frame geometry change. If the layout change is an alternative, it is interesting not to limit the program to satisfy the bay size but to give it a better rating when it does. The function f2 described in Figure A.1 gives the best rating of 1.0 to a perfect fit when c_panel = 1.0. A rating of 0.0 is given when the resulting frame is less than 5 or greater than 15 metres.

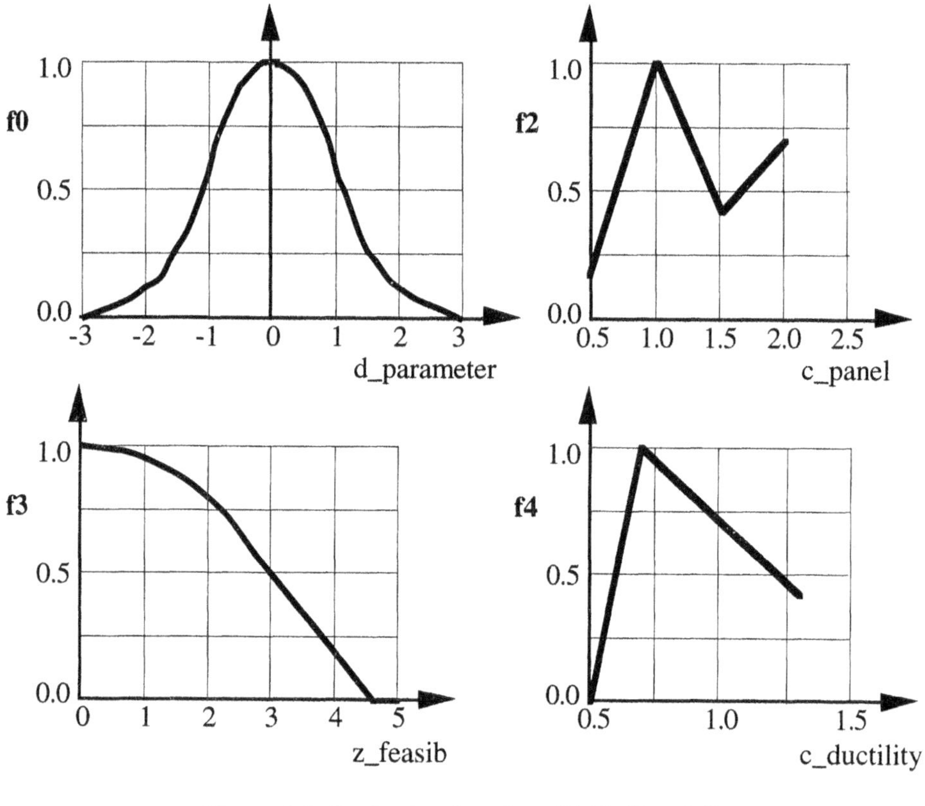

Figure A.1. Evaluation functions f0, f2, f3 and f4.

The evaluation function, f3, measures structural efficiency by considering load resistance through axial forces as more efficient than through bending forces. Designers initially aim to obtain reasonable proportions of a frame so that member selection and detailed verification does not initiate drastic iterations of their design. To obtain reasonable proportions, aspect ratio combinations are used. They are based on one designer's experience and data from existing building frames (Chien, 1988). The aspect ratios of the panel and of the frame provide a good indication of its performance. The performance is divided into zones of feasibility ranging from 0 to 5, as illustrated in Figure A.2.

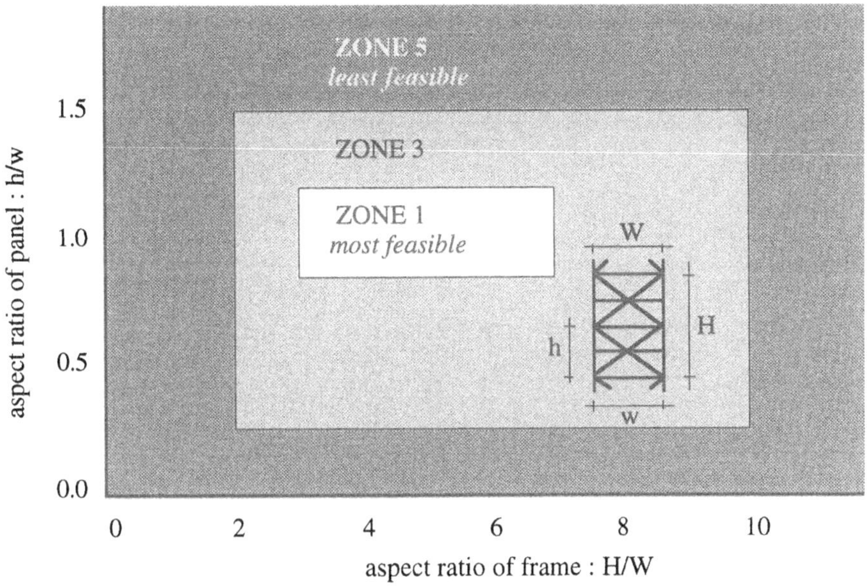

Figure A.2. Zones of feasibility for f3 as a function of aspect ratios

The function f3 described in Figure A.1 gives a better rating for a lower zone. A lower (more feasible) zone is obtained when the frame in question fits aspect ratio guidelines proposed by Chien (1988). The highest (least feasible) zone gets a rating of 0.0, indicating that the phenotype is not very well suited for its design focus. The code below shows the calculation of the value of f3 when the panel type is chevron.

```
if((bldg->type == TY_CHEVRON) &&
     (bldg->mirror == NO) &&
     (((bldg->eccent_1 == EC_NONE) &&
     ((bldg->eccent_2 == EC_W8) || (bldg->eccent_2 == EC_W4))) ||
     (((bldg->eccent_1 == EC_W8) || (bldg->eccent_1 == EC_W4)) &&
     (bldg->eccent_2 == EC_NONE))) &&
     (bldg->m_storeys == 1) &&
     (((bldg->ar_panel > 0.3 ) && (bldg->ar_panel < 0.4)) &&
     ((bldg->ar_frame > 4   ) && (bldg->ar_frame < 10  ))))
  bldg->z_feasib = 1;
```

The evaluation function, f4, assesses structural integrity through a measure of redundancy and plasticity. Designers need to consider the integrity of the frame, in particular for frames subjected to seismic loads, and for large surface structures such as arenas and exposition halls. The configuration of the panel and the eccentricities introduced at the member intersections, provided that adequate measures are taken at the member design level, ensure that multiple load paths and planned plastification will occur instead of hinge formations and member instability. A ductility coefficient is given according to these configuration factors, varying from 0.5 to 1.3, as shown in Figure A.2. A better frame gets a 0.7 coefficient. A less ductile frame gets a coefficient of 1.0 . A coefficient of 0.5 means the frame is too ductile. Such a frame gets a rating of 0.0.

2

EVOLVING BUILDING BLOCKS FOR DESIGN USING GENETIC ENGINEERING: A FORMAL APPROACH

JOHN S. GERO AND VLADIMIR A. KAZAKOV
University of Sydney, Australia

Abstract. This paper presents a formal approach to the evolution of a representation for use in a design process. The approach adopted is based on concepts associated with genetic engineering. An initial set of genes representing elementary building blocks is evolved into a set of complex genes representing targeted building blocks. These targeted building blocks have been evolved because they are more likely to produce designs which exhibit desired characteristics than the commencing elementary building blocks. The targeted building blocks can then be used in a design process. The paper presents a formal evolutionary model of design representations based on genetic algorithms and uses pattern recognition techniques to execute aspects of the genetic engineering. The paper describes how the state space of possible designs changes over time and illustrates the model with an example from the domain of two-dimensional layouts. It concludes with a discussion of style in design.

1. Introduction

There is an increasing understanding of the role that a design language and its representation play in the efficiency and efficacy of any design process which uses that language (Coyne et al., 1990; Gero et al., 1994). A recurring issue is what is the appropriate granularity of a language. If building blocks which constitute the elements of a design map onto a design language then the question becomes what is an appropriate scale for those building blocks in terms of the final design. At one extreme we have parameterised representations where the structure of a design is fixed, all the variables which go to define a design are predefined and what is left is to determine the values of those variables. This defines a very small design space, small in terms of all the possible designs which might be able to be produced for that design situation. At the other extreme we have elementary building blocks which can be combined in a very large variety of ways and which, as a consequence define a very large design space, the vast part of which covers designs

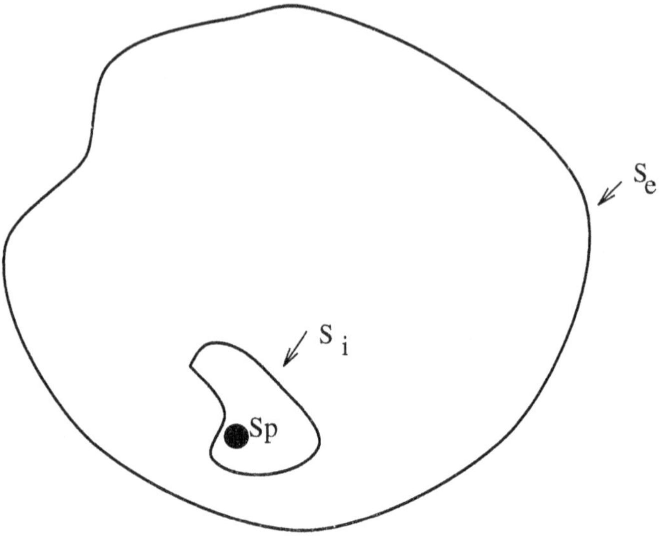

Figure 1. S_e is the design space produced by all the possible combinations of the elementary building blocks, S_p is the design space produced by all the combinations of the values of the parameterised variables, S_i is the design space of interesting designs for the design situation.

which are likely to be uninteresting in terms of the current design situation. The designs produced by the parameterised design representations are a subset of those capable of being produced by the elementary building block representation, Figure 1. Examples of building block representations include constructive systems such as design grammars as exemplified by shape grammars (Stiny, 1980b). Examples of parameterised variable representations include a wide variety of design optimization formulations (Gero, 1985).

The advantage of the use of the elementary building blocks representation is the coverage of the entire design space they provide, whereas the advantage of the parameterised variable representation is the efficiency with which a solution can be reached.

We present here a formal approach which generates a targeted representation of a design problem. A targeted representation is the one which closely maps on to the problem at hand. As an example consider a layout planning problem in architectural design. One representation may be at the material molecular level, where molecules can be combined to make a variety of materials and particular combinations in space produce physical objects; here the potential solution space includes designs which bear no relations to architecture. A targeted representations may be to represent rooms such that the potential solution space primarily includes designs which are all recognizably architectural layouts.

In order to simplify our analysis we consider designs which are assembled from some finite collection of spatial elements (we call them *building blocks* or *compon-*

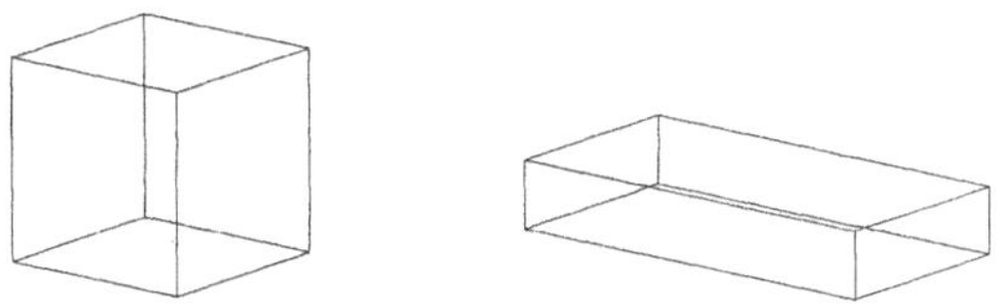

Figure 2. The set of building blocks for Froebel's kindergarten gifts (Stiny, 1980b).

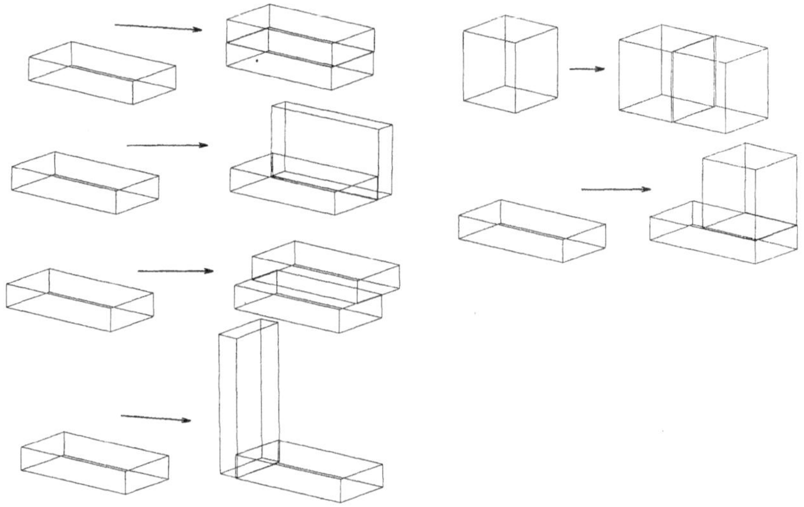

Figure 3. The set of six assembly rules for Froebel's kindergarten gifts.

ents) along with assembly rules. It is assumed that the assembly rules do not affect the components — the design object is a union of non-overlapping building blocks. We start with some set of building blocks which we call *elementary* components. It is assumed that they cannot be decomposed into any smaller ones. We call a set of building components and assembly rules a *representation* of the design problem and the set of elementary components and corresponding rules the *basic* representation. We call it a representation because it implicitly defines the set of all designs (design state space) which can be produced using this set of building blocks and assembly rules.

The kindergarten gifts of Froebel (Stiny, 1980b) is a typical example of such types of design problem. One of many possible elementary representations and assembly rules for it is shown in Figures 2 and 3. One can easily extend it by adding further elementary building blocks and/or further assembly rules.

1.1. TARGETED REPRESENTATIONS

A great variety of designs can be produced within a basic representation. Usually the designer is interested in some particular class of designs. Assume we have some additional set of composite building blocks and an additional set of assembly rules to handle them. We can calculate the number of these composite building blocks which can be found in all possible designs in that particular class and the number of elementary building blocks used to build the rest of these designs (each elementary building block should be counted only once as a member of composite building or elementary building block, the largest composite blocks are counted first and the elementary blocks are counted last). Then we can calculate the frequency of usage of these composite building blocks and elementary building blocks in the entire design space. The same values can be calculated for all designs which have the property or properties we are interested in. If the frequency of the usage of the composite building blocks is much higher for the designs of interest than for all designs built from the elementary building block and the frequency of elementary components usage is much lower than that of the composite building blocks for the design space of interest then we can use the composite building blocks instead of elementary one to produce designs of interest with much higher probability. In other words a representation exists which maps into the area of interest of the entire design space. Let us call it the *targeted* representation for the particular class of designs. Obviously different targeted representations can be produced which correspond to different sets of composite building blocks. We characterize these representations by their "complexity" which is defined recursively as: 0-complexity for the basic representation, 1-complexity for the representation whose building blocks are assembled from elementary building blocks, 2-complexity for the representation whose building blocks are assembled from the building blocks of 0-complexity and 1-complexity, etc. Assume an evolution occurs in an abstract space of complex representations: initially only elementary building blocks exist then a cycle proceeds when a new set of composite building blocks is produced from the ones which are currently available. Then a representation of i-complexity (and building blocks of i-complexity) simply means that composite building blocks of this representation have been produced during i-th step of this evolution.

Different composite building blocks of the same i-complexity may contain different numbers of elementary building blocks: for example, assume some building block of 3-complexity contains 3 elementary building blocks and one of the composite building blocks of 4-complexity is assembled from 2 building blocks of 3-complexity and thus contains 6 elementary components and another one is assembled from one block of 3-complexity and one block of 0-complexity and thus contains 4 elementary components. It is also clear that because there are different ways to assemble the same composite building block it may be produced multiple times in representations of different complexity level during the evolution.

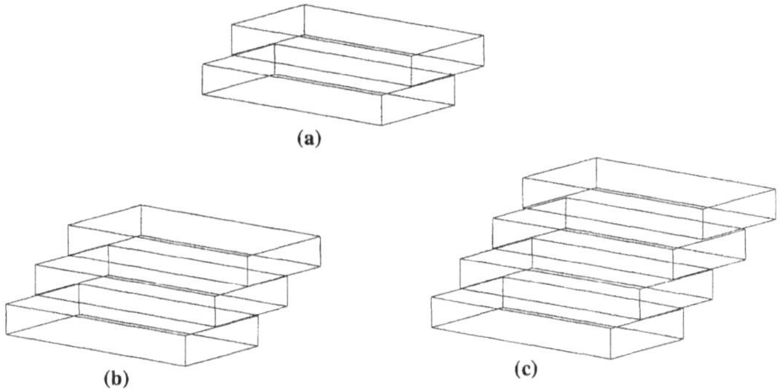

Figure 4. The set of composite building blocks of different complexity for building a staircase; (a) 1-complexity, (b) and (c) 2-complexity.

The search for a reasonably good design using the basic representation is very difficult because significant part of the search effort is wasted in the search of unuseful parts of the design space. If the targeted representation is used instead of elementary one the probability of producing designs of interest becomes much higher, the design space becomes smaller and the design problem less complicated and easier to deal with. The approach presented in this article automatically generates the hierarchy of more and more complex building blocks (in general); ones which are more and more close to the targeted representations which are capable of producing better and better designs.

Assume we work with the representation of the kindergarten blocks shown in Figures 2 and 3 and are trying to design a two-level building with walking access from one floor to the next. The search for a design with this property is quite difficult because only a very small fraction of all feasible objects exhibits it and the probability of discovering the combination of building blocks which makes a staircase during the search is low. However, if we add a composite object of 1-complexity (Figure 4) and corresponding assembly rules Figure 5 to the representation we increase this probability, and if we add a composite building block with 2-complexity (Figure 4) then this probability increases further.

1.2. GENETIC ENGINEERING

Genetic engineering, as used in this paper, is derived from genetic engineering notions related to human intervention in the genetics of natural organisms. In the genetics of natural organisms we distinguish three classes: the genes which go to make the genotype, the phenotype which is the organic expression of genotype, and the fitness of the phenotype in its environment. When there is a unique identifiable fitness which is performing particularly well or particularly badly amongst all the fitness of interest we can hypothesize that there is a unique cause for it and

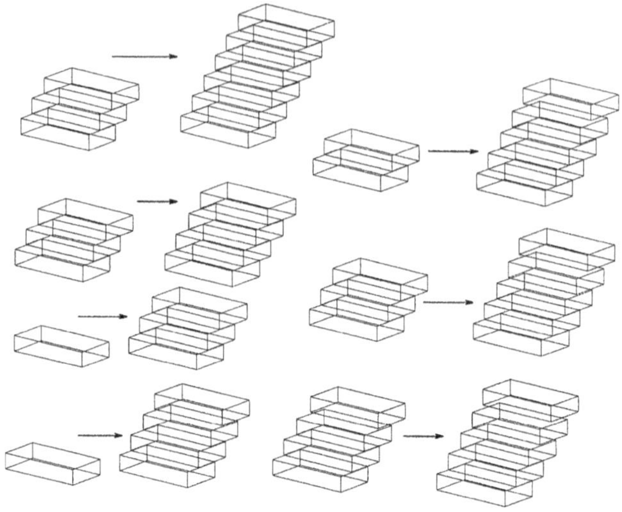

Figure 5. The set of additional assembly rules for handling composite building blocks.

that this unique cause can be directly related to the organism's genes which appear in a structured form in its genotype. Genetic engineering in concerned with locating those genes which produce the fitness under consideration and in modifying those genes in some appropriate manner. This is normally done in a stochastic process where we concentrate on populations rather than on individuals.

Organisms which perform well (or badly) in the fitness of interest are segregated from these organisms which do not exhibit that fitness or do so only in a minimal sense. This bifurcates the population into two groups. The genotypes of the former organisms are analysed to determine whether they exhibit common characteristics which are not exhibited by the organisms in the latter group (Figure 6). If they are disjunctive, these genes are isolated on the basis that they are responsible for the performance of the fitness of interest. In natural genetic engineering these isolated genes are either the putative cause of positive or negative fitness. If negative then they are substituted for by "good" genes which do not generate the negative fitness. If they are associated with positive fitness they are reused in other organisms. It is this later purpose which maps on to our area of interest.

One can interpret the problem of finding the targeted set of building blocks as an analog of the genetic engineering problem: finding the particular combinations of genes (representing elementary building blocks) in genotypes which are responsible for the properties of interest of the designs and regular usage of these gene clusters to produce designs with desired features.

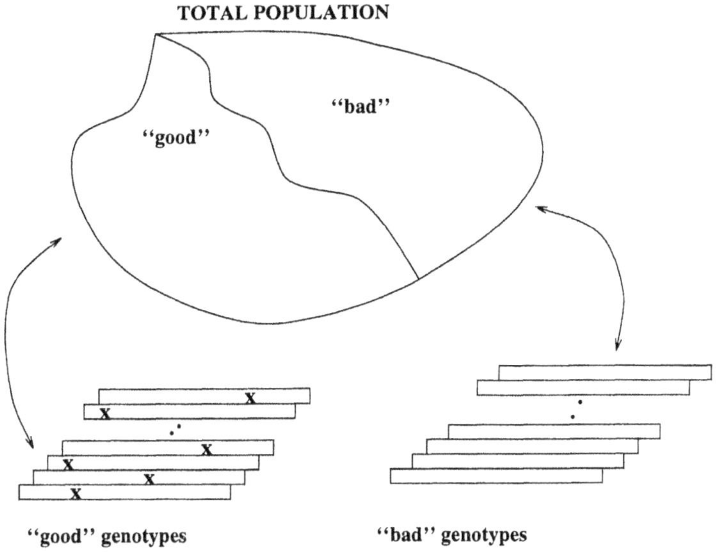

Figure 6. The genotypes of the "good" members of population all exhibit gene combinations, X, which are not exhibited by the genotypes of the "bad" members. These gene combinations are the ones of interest in genetic engineering.

2. Building Blocks

Thus, we establish that different building blocks define different design state spaces (which are, in their turn, the subsets of the entire basic design space). More formally we assume that for the design space of interest a set of composite building blocks exists which is sufficient to build any design of interest from it (or which are sufficient to build a significant part of any of these designs of interest).

We search for these building blocks using the consequence of the assumption made in the introduction about frequencies of composite components usage: on average the sampling set of designs with the desired characteristics (the "good" ones) contains more of such composite building blocks than the sampling set of designs that do not have these characteristics (the "bad" ones). In some cases it is even true in a deterministic sense - that only the designs which can be built completely from some set of composite building blocks possess the objective characteristics, all the rest of the entire basic state space does not have them. One can easily come up with corresponding examples.

In the next section, we describe an evolutionary algorithm which generates "good" and "bad" sampling sets using the current set of building block (set of elementary block at the beginning) and use genetic engineering concepts to determine new composite blocks which are closer to the "targeted" ones than the current set of building blocks. These two steps proceed in cycle while the "good" sampling set converges to the sampling set from the desired design state space and the set of

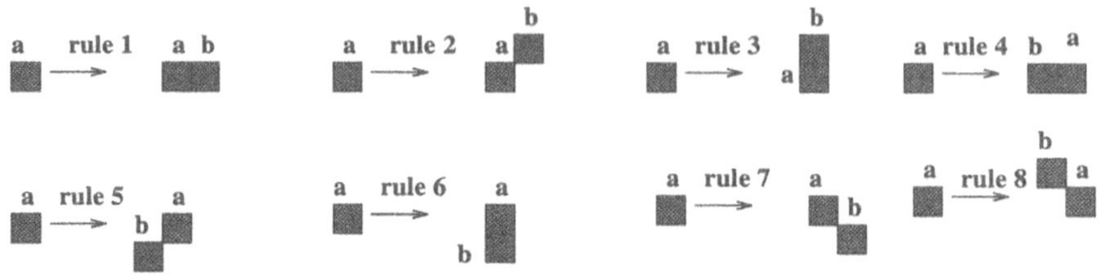

Figure 7. The assembly (transformation) rules used in the example.

complex building blocks comes closer and closer to the targeted set.

If the basic assumption about more frequent use of some composite building blocks to generate the particular class of designs is not true for some problem then the targeted representation for this problem does not exist and the algorithm which is proposed below will not generate an improved representation but will be equivalent to the algorithm for solving the routine design problem (Gero and Kazakov, 1995) and will simply generate the improved designs.

If the sequence of assembly actions is coded as a real vector then the problem of finding the complex building blocks becomes the problem of finding the key patterns in the coding vector — the combinations of codes within it which are likely to be associated with the property of interest in the designs. The vast arsenal of pattern recognition methods can be used to solve this problem. Essentially they are just search methods for subsets in a coding sequence which on average are more frequently observed in objects with desired characteristics than in the rest of the population.

Let us illustrate the execution of the cycle just outlined using a simple 2-dimensional graphical example. We will describe it in more detail later but for now on it is sufficient to say that there is only one elementary block here—the square and that a design is assembled from cubes using the 8 rules shown in Figure 7. Any design can be coded as a sequence of these rules used to assemble it. Assume we are trying to produce a design which has the maximum number of holes in it and that each design contains not more than 20 squares. We start the cycle by generating a set of coding sequences and corresponding designs Figure 8. Then we notice that a number (4) of the designs have the maximal number of holes (designs 1, 2, 4, and 7 - the "good" sampling set) contain the composite building block A and that for three of them their coding sequences contain the pattern $\{2, 8, 5\}$. We also notice that only a few (none in this case) of the designs without holes (designs 3,5,8 and 10 - the "bad" sampling set) contain this block and none contain this pattern in their coding sequence. Then we can generate the next population of coding sequences using the identified sequence $\{2, 8, 5\}$ as a new rule which uses the composite building block A in the design. Assuming that we employ some optimization method to generate this new population we can expect that the "good" sampling set from the new pop-

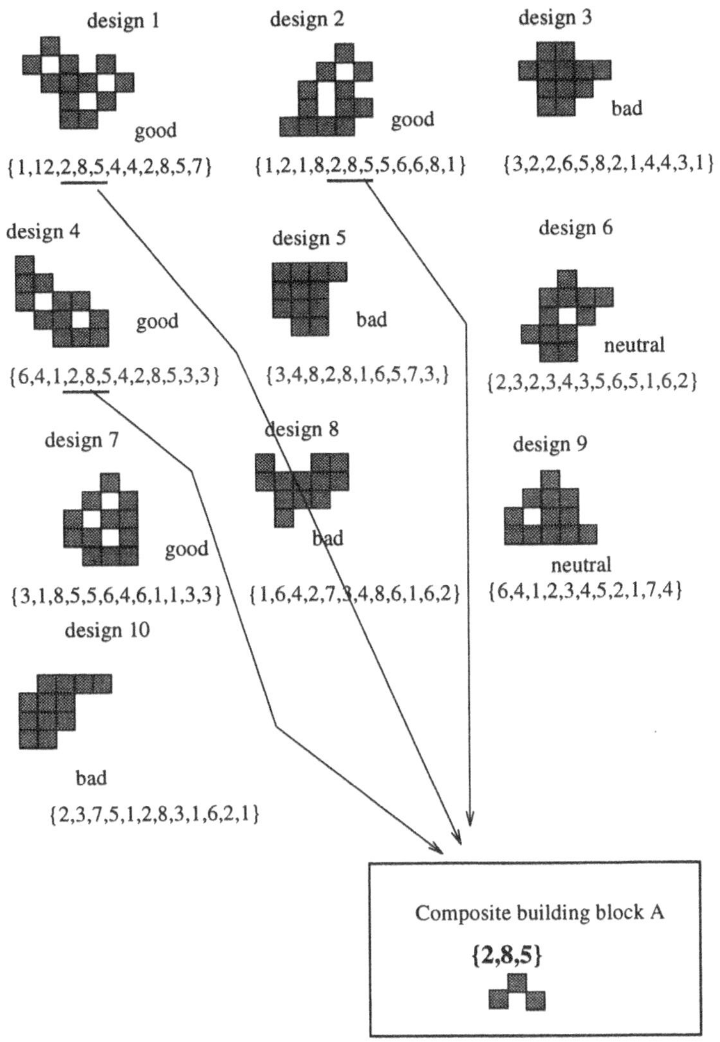

Figure 8. The identification of the pattern $\{2, 8, 5\}$ and corresponding composite building block A in the genotypes of "good" designs.

ulation is better than the previous one (that is, the designs which belong to it have on average more holes than the ones from the previous "good" sampling set). Then we again try to identify the patterns which are more likely to be found in designs from this "good" sampling set than from the "bad" one. This time these patterns may contain the previously identified patters as a component. Then we generate a new population of designs using these additional pattern sequences of rules as an additional assembly rule and so on.

The sizes of the sampling sets in realistic systems is likely to be much larger than the ones in this example and much more sophisticated techniques (Pearson and Miller, 1992) should be employed to single out these key patterns.

3. Evolving Building Blocks

For a more formal analysis of the evolution of the building blocks we use the shape grammar formalism (Stiny, 1980a). It consists of an ordered set of initial shapes and an ordered set of shape transformation rules which are applied recursively. A particular design x within the given grammar is completely defined by a control vector v which defines the initial shape and transformation rules applied at each stage of recursive shape generation. According to the discussion in the Introduction we consider a particular class of shape grammar similar to the kindergarten grammar (Stiny, 1980b), where any shape is a non-overlapping union of building blocks and feasible shape transformations are addition, replacement or deletion of the building blocks.

Let $B = \{b_0, b_1, \ldots, b_n\}$ be a set of n currently available building blocks, and $R = \{r_0, r_1, \ldots, r_m\}$ be a set of m assembly rules applicable to these blocks. Then the control vector $v^i = \{b^i, r^i_1, r^i_2, \ldots, r^i_{N_i}\}, b^i \in B, r^i_j \in R, j = 1, \ldots, N_i$, $i = 1, \ldots, M$ defines the population of M designs $x(v^i)$, $i = 1, \ldots, M$. The length of the control vector $\{v^i\}$, N_i is a variable.

If we add new complex building block $b_{n+1} = x(\{b^{n+1}, r^{n+1}_1, r^{n+1}_2, \ldots, r^{n+1}_k\})$ and new assembly rules $r_{m+1}, \ldots, r_{m+l^{n+1}}$ for its handling then we get a new extended set of rules $R = R \cup_{j=1,l^{n+1}} \{r_{m+j}\}$, $B = B \cup \{b_{n+1}\}$, $n = n + 1$ and $m = m + l^{n+1}$. Now we can produce the design $x(v)$ which corresponds to the vector v whose components belong to the extended B and R. Note that the additional building blocks and assembly rules are generated recursively: they are completely defined in terms of the previous R and B.

We assume that the design problem has a quantifiable objective vector-function $F_k(x)$, $k = 1, \ldots, p$ and can be formulated as optimization problem

$$F(x(v)) = F(v) \to \max \tag{1}$$

The problem (1) over the representation with a fixed set of building components and assemble rules can be solved using any of optimization methods (Gero and Kazakov, 1995) but the stochastic algorithms like genetic algorithms (Holland, 1975) and simulated annealing (Kirkpatrick et al., 1983) look most promising at the moment. We have chosen the genetic algorithm.

The evolutionary method has the following structure :

Algorithm

(0). **Initialization.** Set counter of iteration $k = 0$. Take the set of elementary building blocks $B = \{b_0, \ldots, b_n\}$ and corresponding assembly rules R. Generate some random population of $v^{i,0}$, calculate $x(v^{i,0})$ and $F(x(v^{i,0}))$. Set the relative thresholds for the design's ranking $0 < A_b < A_g < 1$; they are used during an evolution stage to divide the design into "good", "bad" and "neutral" sampling sets, that is, the parts of population which exhibit (A_g) best, (A_b) worse and intermediate relative fitness level.

(1) **Evolution of complex building blocks.** For every component of the objective function F_k divide the population into 3 groups :
"good" ($F_k(x) > \max_{i=1,M} F_k(x_i) - A_g * (\max_{i=1,M} F_k(x_i) - \min_{i=1,M} F_k(x_i))$,
"bad" ($F_k(x) < \min_{i=1,M} F_k(x_i) + A_b * (\max_{i=1,M} F_k(x_i) - \min_{i=1,M} F_k(x_i))$,
and "neutral" (the rest of population).
Determine J combinations $b_{n+j} = x(\{b^j, r_0^j, r_1^j, \ldots, r_{kj}^j\}), j = 1, \ldots, J$ of the current building blocks which distinguish the "good" sampling set from the "bad" one statistically significantly using any one of the pattern recognition algorithms. Add it to the current set of building blocks $B = B \cup_{j=1,J} \{b_{n+j}\}$. Add corresponding new assembly rules to R.

(2) **Generation of new population.** Compute new population using available information about current population $v^{i,k+1} = G((v^{i,k}, x(v^{i,k}), F(x(v^{i,k}))$. The components of $v^{i,k+1}$ belong to the new extended B and R. The G depends on the optimization method employed. If the genetic algorithm has been chosen then $v^{i,k+1}$ is to be calculated using standard crossover and mutation operations. Because the updated grammar includes the grammar from the previous generation the search method guarantees that the new population is better than the previous one (at least no worse) and the new "good" sampling set is closer to sampling set of the design state space of interest.

(3) Repeat steps (1) and (2) until the stop conditions are met.

The stop conditions usually are the termination or slowing down of the improvement in F and/or the end of new building blocks generation.

4. Example

4.1. EVOLVING THE TARGETED REPRESENTATION

As an example we take the problem of the generation of a 2-dimensional block design on a uniform planar grid (derived from Gero and Kazakov (1995)). There is just one elementary component here — a square and the eight assembly rules (transformation rules in terms of a shape grammar) which are shown in Figure 7. If the position where the current assembly rule tries to place the next square is already taken then all the squares along this direction are shifted to allow the placement of new square. It is assumed that the transformation rule at the i-th assembling stage is applied to the elementary block added during the $(i-1)$-th stage. The characteristics of interest are geometric properties of the generated design. In order to demonstrate the idea, assume that the generated design can not consist of more than 32

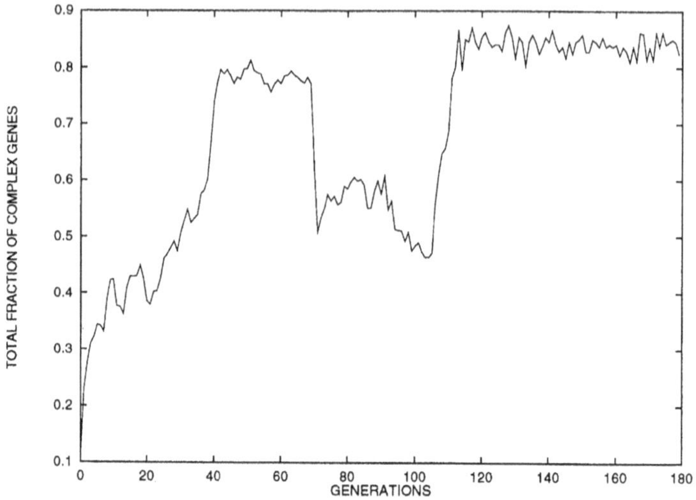

Figure 9. The fraction of composite building blocks in the total pool of building blocks used to assemble the population vs. generation number. The objective function has two components: the area of closed holes and the number of connections between holes and the outside space. The initial set of building blocks contains only elementary building blocks. Evolution proceeds until it naturally dies off.

elementary components. We generate a new population during the stage (2) of the Algorithm using the modification of the simple genetic algorithm tailored to handle multidimensional objective functions (Gero and Kazakov, 1995). We implement a very simple pattern recognition algorithm based on the statistical frequency analyses of double and triple element building blocks with a high cut-off threshold for the acceptance of the patterns. For more complex systems more sophisticated technique is needed.

During the first iteration we begin with the set of building blocks which contains only the elementary ones and search for the designs with maximal area of enclosed holes and maximal number of connections between the holes and outside space. The evolution was allowed to proceed until a stable condition was reached. The result are shown in Figures 9 and 10. By plotting the fraction of the complex building blocks in the total pool of building blocks used to assemble the population at different generations Figure 9, one can see how complex building blocks become dominant and how its fraction reaches a stable level after 110-120 iterations. The fractions of building blocks of different complexity in the total pool at different generation are shown in Figure 10. One can see that during the first 40 generations the total fraction of composite building blocks arises monotonically. For the first 10 generations this rise is completely provided by the increasing number of 1-complexity composite building blocks in the population. Then (from 15 to 30 generations) the fraction of 1-complexity blocks remains stable but the number of 2-complexity building blocks increases and provides the continuing increase

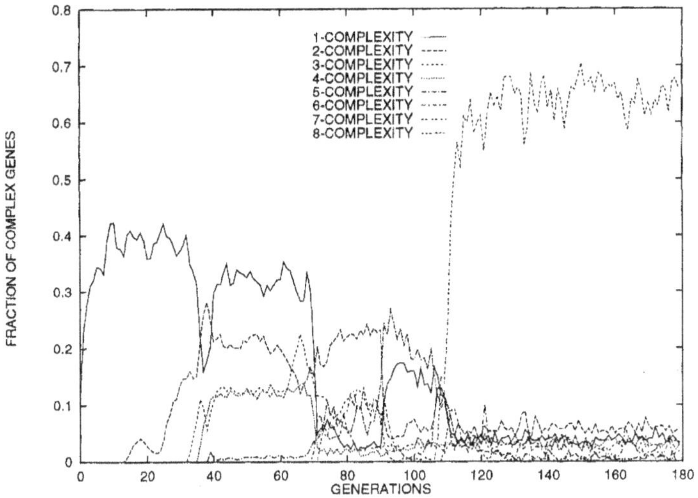

Figure 10. The fraction of composite building blocks with different complexities in the total pool of the building blocks used to assemble the population vs. generation number. This figure shows the building blocks of different complexities which are summed to produce the total fraction shown in Figure 9.

in the total fraction of composite building blocks. From generations 40 to 70 this total fraction is stable with approximately half of building blocks of 1-complexity and half of 2, 3 and 4-complexities. Then the number of 1-complexity blocks and total number of complex blocks declines sharply and from 70 until 110 generation a transitional process occurs with a complex redistribution of populations between representations with different complexities. At the end of this period the building blocks of 8-complexity saturate the population when the fractions of the other complex building blocks are shifted towards a noise level only. One of the evolution paths in the space of complex building blocks is shown in Figure 11 (a). Some of the designs produced are shown in Figure 11 (b). Here arrows show which previously evolved composite building blocks are used to assemble the new building block. The 0-complexity block and its contributions are omitted. As we already noted composite blocks of the same complexity level sometimes have different numbers of elementary components. Coincidently, the 5-complexity block is reproduced again in the representations of 6-, 7- and 8- complexities and is one of the dominant blocks at the end of the evolutionary process.

4.2. USING TARGETED REPRESENTATION

The set of targeted building blocks evolved during this process is then used as an initial set of building blocks during the second experiment when we produce the designs with maximal total area of holes inside and maximal number of connec-

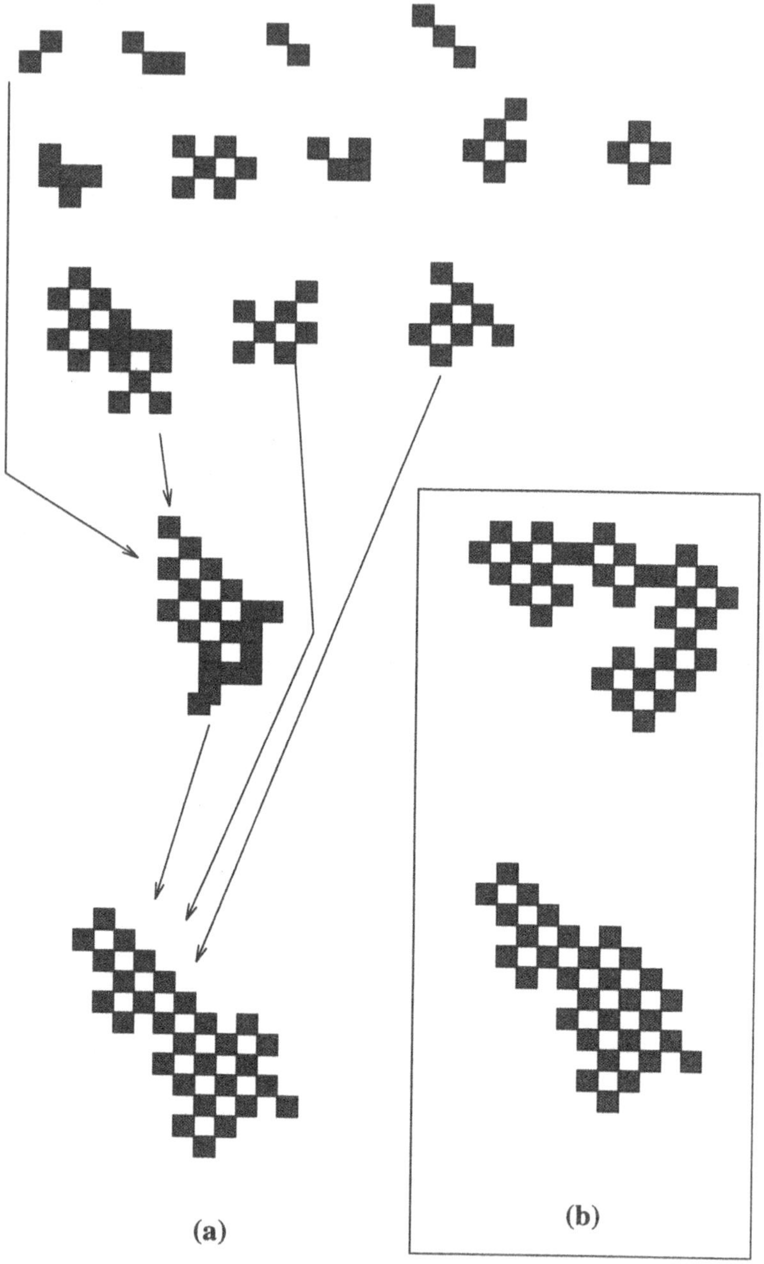

Figure 11. (a) An example of the evolutionary paths in the evolution of a complex building block, (b) some of the designs produced using the set of evolved complex blocks.

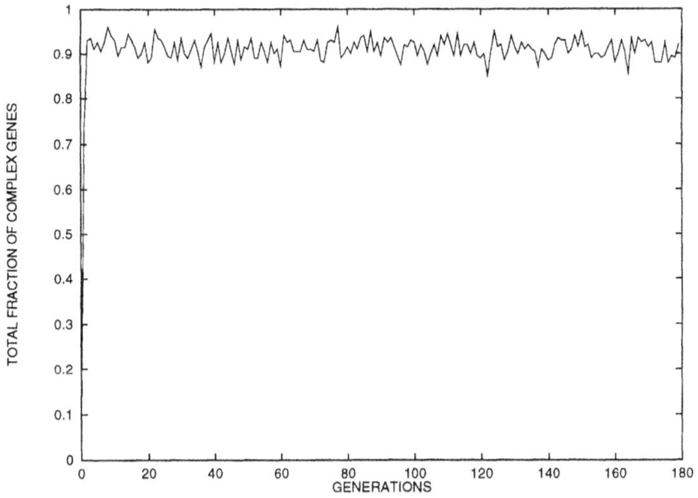

Figure 12. The fraction of composite building block in the total pool of building block used to assemble the population vs. generation number. In this experiment the objective function is the number of closed holes and the number of connection between the closed holes inside the structure. The initial set of building blocks is inherited from the first experiment and is the targeted representation.

tions between these holes inside the structure. Here the fitnesses are close to but not the same as those used to evolve the targeted representation. This experiment is used to test whether the targeted representation is likely to be used more than the original, elementary building blocks. If the targeted representation is used rather than the elementary building blocks then we have achieved our goal of evolving a representation can be used to produce designs which exhibit desired characteristics more readily. The results are shown in Figures 12 and 13. One can see that the fraction of the composite building blocks used to produce these designs reaches the saturation level during the first few iterations. The visible redistributions of the population between the composite building blocks of 5, 6 and 7-complexities are purely superficial - this redistribution occurs between the same composite building blocks which are present in all these representations. Evolution of the representation does not occur during this experiment — no new complex building block were evolved. This can be interpreted as an indication of closeness of the targeted representations for both problems. So if the targeted representation is evolved for one set of objectives then it can be usefully applied to any of the objective sets which are only slightly different to it.

4.3. EFFECTS OF INCOMPLETE EVOLUTION

In this experiment we repeat the first iteration but stop the evolution prematurely after only 60 generations. After this we repeat the second iteration using the evolved

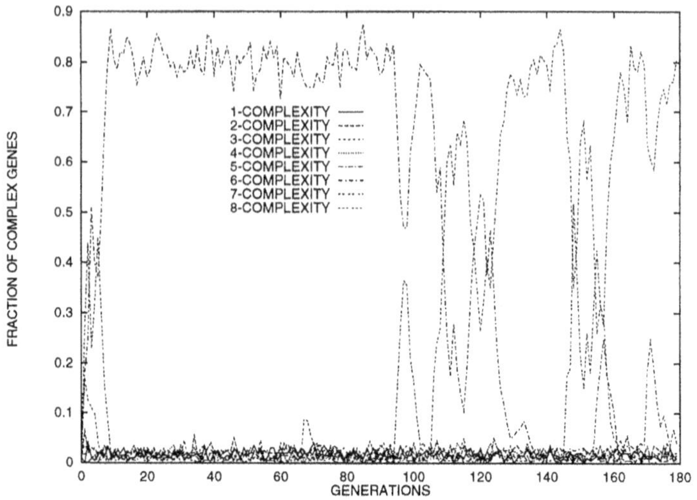

Figure 13. The fraction of composite building blocks with different complexities in the total pool of the building block used to assemble the population vs. generation number in the experiment.

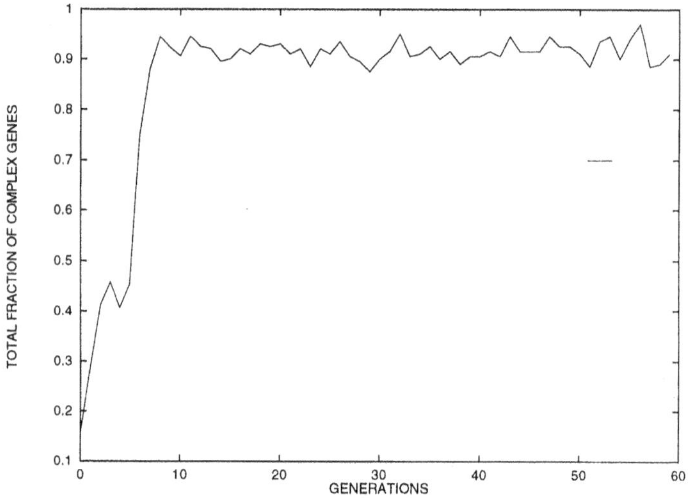

Figure 14. The fraction of composite building block in the total pool of building block used to assemble the population vs. generation number. In this experiment the objective function is the number of closed holes and the number of connections between the closed holes inside the structure. The initial set of building blocks is inherited from first iteration which has been prematurely terminated at generation 60.

incomplete set of composite building blocks. The results are shown in Figures 14 and 15. In this case the evolution of the representation continues for about a further 10 generations and we end up with the same set of evolved composite building blocks. The saturation of the population with the composite building blocks is also

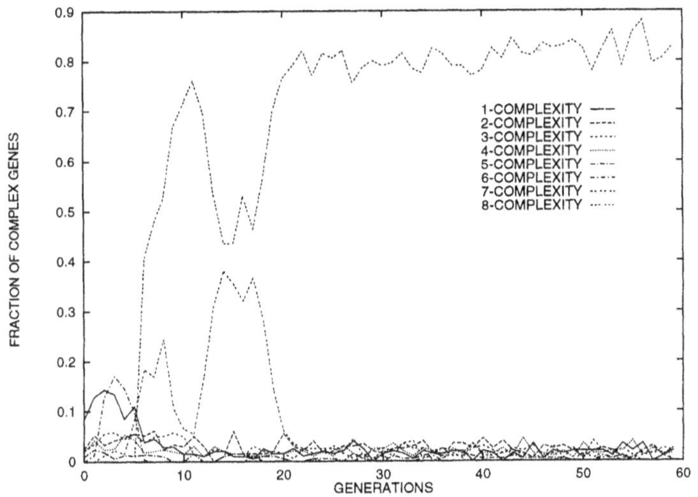

Figure 15. The fraction of composite building blocks with different complexities in the total pool of the building block used to assemble the population vs. generation number in the third experiment.

completed after these 10 generations. Thus, one can start to evolve a representation for one set of objectives and then continue it for another closely related set of objectives.

If we commence by treating the problem as one of finding improved designs then from a computational viewpoint this form of evolution speeds up the convergence to improved designs by up to 15% (in terms of the number of generations required) when compared with standard genetic algorithms. It appears that the use of a targeted representation can lead to the production of designs which are locally optimal.

However, if we use the completion evolution approach presented in the second experiment we get further improvements in performance. We will leave to the Discussion section further discussion of the other advantages of the approach presented.

5. Discussion

The analysis just presented can be easily extended to include general object grammars of types different to the kindergarten grammar. The proposed approach can be considered as an implementation of the simplest version of the genetic engineering approach to the generic design problem. From the technical point of view the algorithm presented is a mixture of a stochastic search method (which may be a genetic algorithm) and a pattern recognition technique.

The genetic engineering approach can be applied in a similar fashion to the problem of the generation of a "suitable" shape grammar (Gero and Kazakov, 1995) where the complex building blocks correspond to the evolved grammar rules.

As already mentioned in the analysis of the numerical experiment, the evolved representations are highly redundant—the same composite building blocks are evolved many times along the different branches of the evolutionary trees. The redundancy level of the current set of composite building blocks can be reduced in a number of different ways. The simplest is just to delete all the redundant copies from the current set. In the general case, we have to find the minimal representation of the subspace which can be generated using the current set of complex building blocks.

The introduction of ideas and methods from genetic engineering into design systems based on genetic algorithms opens up a number of avenues for research into both evolutionary-based design synthesis and into modified genetic algorithms. In design systems based on such modified genetic algorithms it is possible to consider two directions.

The first is to treat the sequence of the genes which results in certain behaviours or fitness performances as a form of 'emergence', emergence of the schema represented by that gene sequence. The use of the genetically engineered complex genes changes the properties over time of the state spaces which are being searched. This allows us to consider the process as being related to design exploration modelled in a closed world. The precise manner in which the probabilities associated with states in the state space change is not yet known. Clearly, this is also a function of whether a fixed length genotype encoding is used or not. If a variable length genotype encoding is used with the genetically engineered complex genes then the shape of the state space remains fixed but the probabilities associated with the states within it change. If a fixed length genotype encoding is used with the genetically engineered complex genes then the shape of the state space changes in addition to the probabilities associated with states in the state space.

The second is to treat the genetically engineered complex genes as a means of developing a representation for potential designs. A fundamental part of designing is the determination of an appropriate representation of the components which are used in the structure (Gero, 1990) of the design. This is part of that aspect of designing called 'formulation', i.e. the determination of the variables, their relationships and the criteria by which resulting designs will be evaluated. In most computer-aided design systems the components map directly on to variables. Further, in such systems the variables are specified at the outset, as a consequence there is an unspecified mapping between the solutions capable of being produced and the variables chosen to represent the ideas which are to be contained in the resulting designs. The genetic engineering approach described provides a means of automating the representation part of the formulation process. The level of granularity is determined by the stability condition of the evolutionary process or can be determined by the user. The targeted building blocks provide a high-level starting point for all later designs which are to exhibit the required characteristics as evidenced in the earlier designs. It is this latter requirement which is met by this formal method.

The following simple picture can be used to summarize the model described in this paper. A group of children are playing with the "Lego" game using not more than 50 squares. They join them together and want to build the object with the largest number of closed spaces inside. After each child has built his or her object the supervisor tries to find a combination of squares which is present in many of the best designs but is present in none or only in a few of unsatisfactory designs. Then he makes this combination permanent by gluing its components together and adds a bunch of such permanent combinations to the pool of building elements available to the children. Then the children make another set of objects using these new building blocks as well as an old ones. The supervisor tries to find another "good" composite block and the process is repeated. Thus, two steps occur in each cycle: first children make a set of new designs from currently available blocks and combination of blocks and second the supervisor tries to single out the additional combination of blocks that should be employed. If there are no such combinations which distinguish "good" design from the "bad" ones then we will not get new combinations but only the improved designs.

Style

The choice of particular variables and configurations of variables is a determinant of the style of the design (Simon, 1975). The label 'style' can be used in at least two ways: either to describe a particular process of designing or as a means of describing a recognizable set of characteristics of a design. Thus, it is possible to talk about the 'Gothic' style in buildings or the 'high tech" style of consumer goods. Precisely what goes to make up each of these styles is extremely difficult to articulate even though we able to recognize each of these styles with very little difficulty. An appropriate question to pose is: how can we understand what produces a style during the formulation stage of a designing process? This brings us back to the concepts described in this paper.

> The history of taste and fashion is the history of preferences, of various acts of choice between different alternatives...... [But] an act of choice is only of symptomatic significance, is expressive of something only if we can really want to treat styles as symptomatic of something else, we cannot do without some theory of the alternatives (Gombrich, quoted from Simon (1975)).

If we use a particular style as the fitness of interest then it should be possible to utilise the genetic engineering approach described in this paper to determine if there is a unique set of genes or gene combinations which is capable of being the progenitors of that style. For this to occur satisfactorily a richer form of pattern recognition will be needed than that alluded to here. We will need to be able to determine a wider variety of gene schemas in the genotypes of those designs which exhibit the desired style.

The use of genetic engineering in evolving schemas of interest opens up a potential subsymbolic model of emergence including the emergence of domain semantics (Gero and Jun, 1995). Style can be considered as a form of domain semantics. This is of particular interest in design synthesis since, if domain semantics can be captured in a form such as described in this paper, then they can be readily used to synthesize designs which exhibit those semantics and even that style. This is analogous to the induction of a shape grammar which captures the characteristics of designer's style.

6. Acknowledgments

This work is directly supported by a grant from Australian Research Council, computing resources are provided through the Key Centre of Design Computing.

References

Coyne, R., Rosenman, M., Radford, A., Balachandran, M. and Gero, J. S.: 1990, *Knowledge-Based Design Systems*, Addison-Wesley, Reading, Mass.

Gero, J. S.: 1990, Design prototypes: a knowledge representation schema for design, *AI Magazine*, **11**(4), 26–36.

Gero, J. S. (ed.): 1985, *Design Optimization*, Academic Press, New York.

Gero, J. S. and Jun, H.: 1995, Emergence of shape semantics of architectural shapes, *Technical Report*, Key Centre of Design Computing, University of Sydney, Australia.

Gero, J. S. and Kazakov, V.: 1995, An exploration-based evolutionary model of a generative design process, *Microcomputers in Civil Engineering* (to appear).

Gero, J. S., Louis, S. and Kundu, S.: 1994, Evolutionary learning of novel grammars for design improvement, *AIEDAM*, **8**(2), 83–94.

Holland, J.: 1975, *Adaptation in Natural and Artificial Systems*, University of Michigan, Ann Arbor.

Kirkpatrick, S., Gelatt, C. G. and Vecchi, M.: 1983, Optimization by simulated annealing, *Science*, **220**(4598), 671–680.

Pearson, W. and Miller, W.: 1992, Dynamic programming algorithms for biological sequence comparison, *Methods in Enzymology*, **210**, 575–601.

Simon, H.: 1975, Style in design, *in* C. Eastman (ed.), *Spatial Synthesis in Computer-Aided Building Design*, Applied Science, London, pp. 287–309.

Stiny, G. : 1980a, Introduction to shape and shape grammars, *Environment and Planning B*, **7**, 343–351.

Stiny, G.: 1980b, Kindergarten grammars: designing with Froebel's building gifts, *Environment and Planning B*, **7**, 409–462.

3

EVOLUTIONARY METHODS IN DESIGN: DISCUSSION

MIHALY LENART
University of Kassel, Germany

AND

MARY LOU MAHER
University of Sydney, Australia

There are numerous approaches to modeling or describing the design process by using formal methods. Different approaches are taken for tackling different tasks or for modeling different aspects of the design process. For example, formal grammars are used for capturing generative tasks or knowledge-based and case-based reasoning for utilizing domain knowledge or expertise in order to find solutions in a large and complex search space. In recent years evolutionary methods became popular in engineering and computer science for solving hard, and with other methods sometimes intractable problems. They have proven to be successful in such areas where there is little domain knowledge available and the solution needs not to be perfect, just good enough for the particular task.

Research on evolutionary methods started in the 1950s and the first computational models were developed in the 1960s by I. Rechenberg, H. P. Schwefel and L. Fogel. A significant development came with J. Holland's and K. De Jong's work on Genetic Algorithms (GAs) in the 1970s. Since the early 1980s there has been an explosion of research and the number of new applications in science and engineering has been growing exponentially. In design, evolutionary methods have first been applied to optimization problems. More recently, they are also used for solving design problems which require creativity. The main concern of the papers presented at this workshop and the subsequent discussion was the role of evolutionary methods in design exploration, in particular, how evolutionary methods can support human creativity in the design process.

1. Papers

This session included two paper presentations:

1. *Modelling Design Exploration as Co-Evolution: A Combined Gene Approach* by Mary Lou Maher, Josiah Poon and Sylvie Boulanger.

2. *Evolving Building Blocks for Design Using Genetic Engineering: A Formal Approach* by John S. Gero and Vladimir A. Kazakov.

The presentations showed two different approaches to the use of GAs in formalizing aspects of the design process. The first presentation was concerned with how GAs provide a mechanism for change over time and applied this mechanism in a way that allowed the design focus to change as well as the design solution. This was modeled as a co-evolutionary process in which two spaces co-evolve: the *problem definition space* and the *solution space*. The presentation emphasized the focus on exploration rather than on search that is characteristic for other computer based design models. The co-evolutionary paradigm has been chosen for it is less deterministic than traditional evolutionary paradigms. Problem solutions and definitions cooperate in order to improve their fitness. Fitness is defined, however, only locally which means that the fitness function, i.e., the focus of the design process might change as new generations evolve.

The second presentation was concerned with how GAs provide a basis for identifying representations that achieve a high level of performance. This mechanism is applied in a manner that is analogous to genetic engineering in which the representation of design genes evolved. The approach allowed the representation of design knowledge to change in response to good performance. It generates building blocks of a design language, starting from simple blocks from which successively more and more complex ones will be generated during the design process. Analogous to genetic engineering, emerging patterns are recognized and used for subsequent evaluation. The evaluation also identifies bad genes in which case either the proliferation of bad genes is cut off by halting the generation process or the bad genes are "treated" in order to get rid of undesirable features.

2. Discussion

The issues raised during the discussion were directed at one or the other presentation and are presented here according to the presentation that raised the particular issue.

2.1. CO-EVOLUTION ISSUES

Three major issues were raised:

1. *Should the problem definition change or should it be refined during co-evolution?*
   This discussion raises a basic question of whether formal methods developed for solving a particular design problem should be allowed to change the design problem in the course of the process or should the formal methods be refining and adding to the problem definition. In other words, the question is how to define design exploration; where are the limits for changing the problem

definition and at which abstraction (meta) level we want to carry out such changes.
The discussion introduced a few scenarios in which the problem definition would be changed and whether the problem should have been restated at a higher abstraction level rather than changed at a lower level. In particular, the design of a bridge was discussed where the initial design may prove to be intractable and the designer may change his mind and choose to design a tunnel crossing. This, as an example of a change in problem definition, could also be considered a refinement of a problem definition of designing a harbor crossing.

2. *Do the dominant criteria eliminate consideration of other criteria?*
This discussion raises the issue of loss of information when using GAs. When allowing the problem definition as a set of criteria to evolve according to a fitness function, it is possible that some initial criteria may be lost due to poor performance. In the model of design exploration as co-evolution, this potential loss may mean that the formal method does not solve the initial problem at all but solves some evolved representation of the problem. This problem is related to the lack of global fitness and thus to the possibility of changing focus in the course of the design process. There was some concern as to whether this was an appropriate model for CAD.

3. *Do such formal methods need to model a global set of design criteria, or is it reasonable to model the fitness as a set of local criteria?*
This discussion dominated the discussion period by raising an issue that could not be resolved. The model of co-evolution that uses the combined gene approach relies on a local fitness function that is defined within a genotype. This local fitness function provides the criteria for evaluating the individual genotype and there is no model of a global fitness. The global fitness introduced in the paper was used to track progress but was not used to guide the evolutionary process.
The arguments for a *global fitness* include: a set of criteria by which all design solutions are evaluated means that there is a common unit of measure for fitness; global criteria means that there is a representation of the problem being solved that is related to the context rather than the solution; a global fitness can directly be compared to the original problem as defined by the user's initial requirements. The lack of global fitness means also that the process does not converge and after a certain number of steps there is no way to tell what we have accomplished so far and where to stop.
The arguments for a *local fitness* include: each design solution should be evaluated on its own merit rather than on a predefined ruler; a local fitness allows other criteria to be introduced based on the solutions being considered; local criteria need not relate to the original problem definition since the initial re-

quirements for a design problem may not be an appropriate characterisation of the need.

2.2. GENETIC ENGINEERING ISSUES

Three major issues were raised:

1. *Is the correlation between good genes and performance causal or statistical?* The concern in raising this issue is in the use of the genetic engineering approach to develop representations of design knowledge that are used as a basis for new designs. If the genes are evolved according to performance without an understanding how the performance relates to the genes in a causal way, then new representations are statistically good. A lack of causal understanding may cause the evolution to follow the limited measure of performance. The discussion was resolved that the presented method did rely on a statistical correlation and that this was as good a correlation as any other.
2. *Are the evolved genes specialised for a specific criteria?* Given that the new genes evolved in response to performance of a specified criteria the implication is that these are specialised genes. This lead to a similar, related issue of whether the evolved genes could be used for other purposes. A flexibility of representation would be sacrificed because the method relies on a fixed performance measure.
3. *Can a bad gene be used again?* This issue is similar to the loss of information issue raised in Section 2.1. If the bad genes are discouraged from being used, does this result in a loss of information in the genotype representation? The discussion brought out the details of what is meant by a bad gene. In fact, genotype representations are not modified, but certain combinations are encouraged and others are discouraged. This means that theoretically there is no loss of information. Practically there may be a loss if certain genotypes are discouraged to a point where they are not used at all. However, if the performance criteria changes, these bad genes may start to reappear again.

3. Summary

The discussion could be summarised as two statements and two questions:

Statement 1: The use of GAs presented and discussed in this session employs a mechanism to change the design space: the co-evolution methods change the representation of the design requirements and design solutions, and the genetic engineering method changes the representation of design knowledge.

Statement 2: Evolutionary methods provide a set of mechanisms for generating design solutions and rely on evaluation knowledge to guide the design process.

Question 1: How do we know if the changing design space is still relevant? The co-evolution methods change the design requirements to a point where they may not have anything in common with the original requirements. The genetic engineering method changes the design knowledge to a point where the original design knowledge representation may never be used at all.

Question 2: Do these evolutionary methods do something fundamentally different to other formal methods of design? What kind of design problems can be tackled with evolutionary methods better than with other formal methods?

The answers to these two questions point to new research directions for formal design methods based on evolutionary approaches.

PART TWO

Generative and Search Methods in Design

4

MODIFIED SHAPE ANNEALING FOR OPTIMALLY-DIRECTED GENERATION: INITIAL RESULTS

KEN N. BROWN[§] AND JON CAGAN
Carnegie Mellon University, USA

Abstract. Formal generative design systems may describe large and complex spaces, making good designs hard to find. The problem of directing generation towards good solutions is addressed by considering the shape annealing algorithm. The algorithm is modified in an attempt to make optimization over the final generation sequence more uniform. Seven variations of the algorithm are investigated, and their performance compared. The results suggest that an enhanced backtracking ability may significantly improve the performance of the algorithm.

1. Introduction

Formal generative systems promise a number of advantages for a science of design. They provide an executable specification of a design space, supporting the generation of designs within the space. They offer, in principle, an analyzable definition of the design space, allowing candidate designs to be recognized as elements of the space. They provide (again in principle) a means to analyze properties of the spaces themselves. They offer a formal basis for reasoning about designs in progress, via their semantics. In order to meet these promises, a number of questions have to be answered.

Although formal systems can be concise and elegant descriptions of design spaces, the spaces themselves can be vast and the relationship between the description and the designs its specifies unclear. The result of applying sequences of individual rules is not obvious. How is a designer to know what designs exist in the space, and which regions of the space contain what types of designs? How is a designer to find good designs within the space? How should the syntax of a system be interpreted, and in particular, how should partial, incomplete designs in the space be interpreted? Intelligent generation within a space would seem to require either a knowledge of the space's overall structure, or some method of judging intermediate states and making

---

[§] Now at University of Aberdeen, Scotland.

decisions about which directions to follow. More complete discussions of these issues are given by Carlson (1994b) and Cagan (1994b).

One technique that has been proposed for searching large design spaces is *shape annealing* (Cagan and Mitchell, 1993), a search method directed towards optimal solutions, based on the stochastic optimization technique of simulated annealing (Kirkpatrick et al., 1983). Shape annealing is useful in cases where the intermediate assessments are such that simply applying the locally best operation at each stage does not necessarily lead to the best final designs, as its stochastic nature allows it to backtrack out of local optima. Shape annealing produces good solutions efficiently; however, it is not guaranteed to find the optimum, and seems to find it only rarely. In this paper, we consider why this should be so, and suggest some modifications to the basic technique in order to make reaching the optimum more likely.

2. Background

The use of grammars (formal generative systems) for the design of complex structures was popularized by the development of the shape grammar formalism (Stiny, 1980). In this formalism, rewrite-rules are recursively applied to two-dimensional shapes to produce languages of two-dimensional shape designs. Parametric shape grammars can also be defined, in which rule schemas with variable parameters are specified, and rule application proceeds with an instantiated version of one of the rule schemas. A number of shape grammars have been presented in the literature, including a grammar of Palladian villas (Stiny and Mitchell, 1978) and a grammar of Queen Anne houses (Flemming, 1986). A general definition of the algebra of shapes is given by Stiny (1991). Grammars were originally used as a means of analyzing and understanding existing corpuses of designs, by showing that the designs conform to a regular structure.

Viewed in the abstract, the whole design process can be considered to be generation. Stiny and March (1981) proposed *design machines*, while Fitzhorn (1989) has proposed a formal computational theory of design. In both of these papers, the design process is modeled as the interaction of constraints and the design context with the grammar rules used to generate the designs. Brown et al. (1993b) discuss some possible roles for grammatical methods in engineering design, in the context of a transformational model of design (McMahon et al., 1993). Recently, there has been some effort in building grammatically-based design tools, allowing designers to explore design spaces interactively (Carlson, 1994a; Heisserman and Woodbury, 1994).

The use of grammars as design tools raises the question of how designs should be interpreted. The conventional view of the semantics of a design representation provides a mapping from the elements of the representation to real world objects. We can also look at performance criteria, or objectives, as the semantics of a design, either in terms of rigorous mathematical notions of performance, or the more fluid and subjective assessments of human users. We can use the constructive nature of grammatical tools to construct these semantics in tandem with the syntactic generation. Stiny (1981) proposed the generation of design descriptions by associating rules operating on descriptions with the grammar rules which operate on the shapes. Brown et al. (1994a, 1994b) have applied variants of this formalism to the interpretation of engineering artifacts in terms of feature descriptions and manufacturing plans. For string and graph based formalisms, attribute grammars (Knuth, 1968) have been used, in which attributes representing additional information augment symbols, and attribute rules define the computation of the attribute values. Variations on this theme are proposed by Penjam (1990), representing the resistance of electrical circuits, Rinderle (1991), representing forces and weights of boom designs, and Brown et al. (1992), representing stress concentrations on loaded shafts. However, these applications are oriented towards providing a final interpretation or assessment of a design. When used in design tools, the assumption is that the partial interpretation constructed as a design is in progress can be used to direct generation towards a desired end result. This is not necessarily the case, and depends on the particular semantic functions being used. The problem remains for each particular design space and design task of determining objective functions with which variations in the intermediate assessments after individual rule applications accurately reflect the variations in the assessment of the final designs resulting from those rules.

The question then arises of how best to guide search or exploration through these design spaces, which are generally too large for exhaustive search to be practical. For exploration, random sampling of the space may suffice, although as Carlson (1994b) points out, for a truly uniform sample of the space, the frequency with which particular rules are applied in particular situations should depend on the size of the sub-space below the states that result. For search in relatively small spaces, standard artificial intelligence search algorithms may be applied. For larger spaces, Cagan and Mitchell (1993) proposed shape annealing as a means of directing search towards optimal solutions. Shape annealing attempts moves more or less at random which are accepted according to a steadily more discriminating acceptance criterion. Thus, a search in a design space consists of successive rule applications and retractions which are initially randomly selected and accepted almost regardless of their effect on the objective function. As time progresses, moves which degrade the objective function are accepted with decreasing probability. Eventually, the algorithm converges to a good local optimum. Because shape annealing is stochastic, it is able to recover from early, poor local optima. Thus, intermediate objective functions do not need to be strictly accurate, as moves which initially appear to take us closest to our goal but later turn out to have been non-optimal are not fatal to the algorithm. However, the choice of intermediate objective function is still crucial to the algorithm's success. Shape annealing has been applied to component layout (Szykman and Cagan, 1994), the generation of truss topologies (Reddy and Cagan, 1994), and geometric knapsack problems (Cagan 1994a).

As stated earlier, although shape annealing is an optimally-directed search technique, and produces good solutions, the final designs are rarely the global optimum, and in many cases are far from it. In the following section, we discuss the algorithm in more detail, and consider why this might be the case.

3. Shape Annealing

Let G = (S, L, R, I) be a shape grammar, consisting of a set of shapes, S, a set of labels, L, a set of shape rules, R, and an initial labeled shape I. Let $f:(S,L)^+ \rightarrow \mathbf{R}$ be the objective function, mapping labeled shapes to real numbers. Let T be a real-valued variable, called the temperature.

A state s_i is a labeled shape obtainable by recursively applying a sequence of n rules from R to I ($n \geq 0$). Let $f(s_i) = C_i$ A shape annealing move is then the application of a grammar rule r_j to s_i or the retraction of the last rule applied to obtain s_i, to obtain the labeled shape s_{i+1}. Let $f(s_{i+1}) = C_{i+1}$. If $C_{i+1} < C_i$, then s_{i+1} is accepted as the next state. If $C_{i+1} > C_i$, then s_{i+1} is accepted as the next state with a probability defined by

$$\Pr(\text{accepting } s_{i+1}) = e^{-|(C_{i+1} - C_i)/T*Z(T)|}$$

where Z(T) is a normalization factor. T decreases with time.

Although it can be proven that simulated annealing under certain conditions relating to the parameters of the algorithm will always find the global optimum, the conditions are unfeasibly restrictive for real applications. At least the same restrictions apply to shape annealing. However, it appears that when tested on the same problem, shape annealing can produce inferior solutions to standard simulated annealing (Szykman and Cagan, 1994). There are two main reasons for this. Firstly, as stated above, the choice of objective function for intermediate designs is significant. It appears to be harder to relate the early stages of a design to the final design and make decisions on that basis than it does to represent violations of constraints in fully instantiated designs and minimize those violations. Secondly, there is a subtle difference in the implications of accepting a move in the two algorithms. This point is discussed below.

In the original shape annealing application of Cagan and Mitchell (1993), a design is produced by a sequence of moves, each one of which constrains the subspace that is subsequently reachable. Each step in a generation can thus be said to contribute to the final cost of the design by imposing constraints on the costs that can be reached. After accepting a particular move, in order to generate a design in the subspace that is outside the resulting subspace, the shape annealing algorithm has to backtrack over the moves responsible. The length of the move sequence generally grows with time. However, the acceptance criterion becomes more discriminating with time as the temperature drops. Thus decisions on whether rule applications and retractions are accepted are not uniform over the length of the sequence, and optimization is concentrated towards the end. If we had complete knowledge of the design space and the objective function (i.e. we could characterize completely the relationships between the states in the space and the objective functions of the final designs below them) we would see that the optimality of the individual moves increases with the length of the sequence. This phenomenon was referred to implicitly in the original paper by Cagan and Mitchell (1993), in which they discussed the way in which the results are crucially dependent on the early choice of moves. In other words, if we are fortunate in our initial choices, we end up with a solution close to the optimum; if we are unfortunate, then no amount of optimization at the end of the sequence will produce the optimum. (This is not to say that the early moves in the sequence are randomly chosen, as the algorithm frequently backtracks over those moves and tries alternatives. It is simply that when the algorithm is concentrating on the early stages, the acceptance criteria is loose, and as the criteria becomes more stringent and the sequence lengthens, the algorithm is less able to backtrack long distances). In addition, the algorithm has no memory of the states it has visited, and once it has backtracked out of a superior path, it has no knowledge of the previous objective functions it had discovered and may subsequently settle on poorer states. Finally, it should be noted that shape annealing in itself is not the cause of the above behavior, but its application to generative systems in which each move narrows the space of reachable designs. These arguments do not apply to systems in which any state can be reached from almost any other state by forward rule applications — for example, Reddy and Cagan's (1994) truss topology grammar. In this paper, we restrict our comments to those systems in which a forward move does limit our space of designs.

These comments suggest a number of possibilities for improving the algorithm. One is to ensure that the objective functions at the early stages better reflect the utility of the sub-spaces that the moves create, either by finding a better objective function, or by changing the grammar so that the initial moves create better partitions of the whole space. A second is to implement some form of memory, so that previous objective function evaluations are incorporated in some way into the algorithm. A third to is change the move set available to the algorithm to try to compensate for the non-uniform optimization. Cagan and Kotovsky (1994) have investigated the propagation of objective function evaluations from states to neighboring states, effectively combining the memory approach with the improved objective function approach. Schmidt and Cagan (1994) have considered *recursive* annealing, in which design progresses through different levels of abstraction, annealing at each stage, and thus providing better estimates of subsequent costs when deciding upon moves at higher levels. For the remainder of this paper, we will consider the third option, and investigate alternative move sets.

4. Experiments

In the experiments described below, we had two main aims: to allow the acceptance criterion to use information about the costs of subsequent states in the algorithm, and to increase the likelihood of the algorithm backtracking out of deep local optima. We compared seven different backtracking strategies, sketched in Figure 1, and described below. In what follows, we will refer to the current state as s_{old}, and the state the algorithm is attempting to move to as s_{new}.

s_{old} is a state obtained by applying rules $(r_{i_1}, r_{i_2}, ..., r_{i_n})$ to the initial shape. Let C_{old} be the old evaluated cost and C_{new} be the new evaluated cost used in the probability calculation.

A. s_{new} is the state obtained by applying a rule $r_{i_{n+1}}$ to s_{old}, or by retracting r_{i_n}. $C_{old} = f(s_{old})$, $C_{new} = f(s_{new})$.

B. s_{new} is the state obtained by applying a rule $r_{i_{n+1}}$ to s_{old}, or by retracting rules r_{i_n} back to r_{i_j}, for $1 \leq j \leq n$. $C_{old} = f(s_{old})$, $C_{new} = f(s_{new})$.

C. s_{new} is the state obtained by applying a rule $r_{i_{n+1}}$ to s_{old}, or by retracting rules r_{i_n} back to r_{i_j} to get state s_j, followed by applying a rule r_{i_j}' to s_j. $C_{old} = f(r_{i_j}(s_j))$, $C_{new} = f(s_{new})$.

D. s_{new} is the state obtained by applying a rule $r_{i_{n+1}}$ to s_{old}, or by retracting rules r_{i_n} back to r_{i_j} to get state s_j, followed by applying a rule r_{i_j}' to s_j. $C_{old} = f(s_{old})$, $C_{new} = f(s_{new})$.

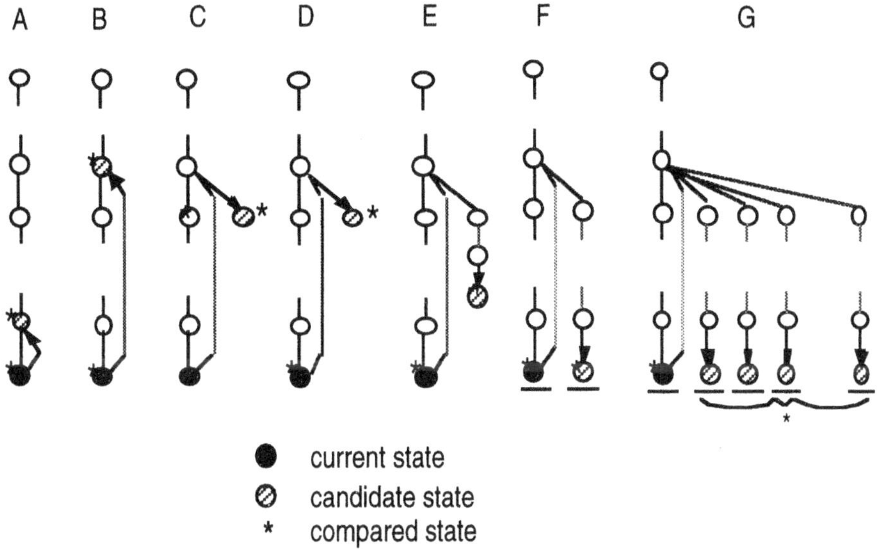

Figure 1. Alternative backtracking moves.

E. s_{new} is the state obtained by applying a rule $r_{i_{n+1}}$ to s_{old}, or by retracting rules r_{i_n} back to r_{i_j} to get state s_j, followed by applying rules r_{i_j}', ..., r_{i_m}', such that m-j = min(n-j, x) for some fixed x. $C_{old} = f(s_{old})$, $C_{new} = f(s_{new})$.

F. s_{old} is a finished design. s_{new} is the finished design obtained by retracting rules r_{i_n} back to r_{i_j}, for $1 \le j \le n$, followed by applying rules r_{i_j}', ..., r_{i_m}'. $C_{old} = f(s_{old})$, $C_{new} = f(s_{new})$.

G. s_{old} is a finished design. s_{new_k} for all k such that $1 \le k \le p$, for some fixed p, is the finished design obtained by retracting rules r_{i_n} back to r_{i_j}, for $1 \le j \le n$, followed by applying rules r_{k_j}', ..., $r_{k_{m_k}}'$. s_{new} is a state s_{new_t} such that $f(s_{new_t}) = \min(f(s_{new_k}))$ for all k: $1 \le k \le p$. $C_{old} = f(s_{old})$, $C_{new} = f(s_{new})$.

Move set A is the original shape annealing algorithm. B is intended to make it easier to back out of local optimum, by allowing the retraction of an arbitrary number of the applied rules. Although the probability of acceptance will be lower than for any single backtrack step, it should be higher than the accumulated probability of accepting the equivalent sequence of single backtracking steps. For C and D, it was noted that backtracking involves the loss of useful information. Suppose the transition from s_i to s_{i+1} was a good transition, but that we backtrack to state s_i. The algorithm will now accept any move that results in a better state than s_i, even though it may be much worse than s_{i+1}. We need some way of balancing the ability to backtrack out of local optima with the retention of information from previous moves. Thus for C, a backtracking move consists of the retraction of a number of rules to state s_i, followed by a forward rule application to give a new state s_{i+1}'. Comparison of costs for the acceptance criterion is then between s_{i+1} and s_{i+1}'. D is the same as C, except that comparison is between s_n and s_{i+1}'. E extends this "look forward" idea by applying a number of forward rule applications (up to a limit) before comparison. F is motivated by an attempt to use as good an estimate of the final objective functions as possible. The best possible estimate is the objective function for finished designs — therefore, in F, comparisons are only made between completed designs. This algorithm first randomly applies rules until a completed design is obtained; an annealing move then consists of a backtrack to state s_i, followed by a new random completion from s_i to s_m'. The objective function values for s_n and s_m' are then compared as normal. Finally, G is based on F, except that after a backtrack, a fixed sample of completions are generated, and the best completion is used for comparison.

In order to test the various moves sets, we used a simple generative system as a test problem. The problem was devised to have a seemingly obvious optimum move at every stage, but with a small number of local optima. The system generates descending lists of real numbers from 100 to 0. Given a partial list, a move consists of simply selecting the next element. Generally, a move involves subtracting a number between 1 and 5 (in integer multiples of 0.01) from the current last element, X_i, to obtain the next element X_{i+1}, such that $X_{i+1} \geq 0$. However, there are a few exceptions. If X_i is between 61 and 62, then $X_{i+1} = 55$. If X_i is between 60 and 61, the $X_{i+1} = 59.99$. If $X_i = 59.99$, then $X_{i+1} = 59.98$. Corresponding exceptions apply to values in the range 29.99 to 32. A more formal description of the system is given in the appendix.

Our design task will be to generate short lists, and thus the objective function for a completed list is simply its length minus 1. The optimum cost is 20, and there are many lists with that cost, for example,

<100 95 90 85 80 75 70 65 61.5 55 50 45 40 35 31.2 25 20 15 10 5 0>.

Note that without knowledge of the structure of the space, the best move at any time would appear to be to subtract the largest number possible from X_i. However, in certain cases (where this would leave us in the ranges [59.99,60.99] or [29.99,30.99]) the best move is actually to subtract a smaller number. Thus a straightforward hill-climbing search with a naive cost estimate, which always selects the best local move, will not find the optimum cost.

In addition to the different move sets, we also implemented a number of different intermediate objective functions, to see if there was a relationship between objective function and move set. The cost functions are described below.

Let L be a partial list of length N with last element i. Each cost function uses an estimate of the number of moves required to complete the list, by assuming an average decrement over all the moves. The "max(1000*i,0)" function is to reward moves which complete the list when $i < 1$, and the $1/N$ factor in function 5 encourages forward generation.

1. $f(L) = N + (i * 100) + \max(1000 * i, 1)$
   ("worst cost" — assumes an average decrement of 0.01).

2. $f(L) = N + i + \max(1000 * i, 1)$
   ("poor cost" — assumes an average decrement of 1).

3. $f(L) = N + i/2.5 + \max(1000 * i, 1)$
   ("average cost" — assumes an average decrement of 2.5).

4. $f(L) = N + i/4 + \max(1000 * i, 1)$
   ("good cost" — assumes an average decrement of 4).

5. $f(L) = N + i/5 + \max(1000 * i, 1) + 1/N$
   ("best cost" — assumes an average decrement of 5).

In running these experiments, the adaptive annealing schedule of Huang et al. (1986) was used, which calculates initial temperatures, temperature decrements and equilibrium criteria by statistical sampling of the algorithm's performance. For determining the size of a backtracking move, we considered five percentage bands (0-20%, 20-40%, etc.). First we select a percentage band, compute the limits on backtracking that imposes, and then randomly select a backtrack height within those limits. To select the percentage band, we use the probabilistic move selection technique suggested by Hustin and Sangiovanni-Vincentelli (1987). This technique computes the average objective function change induced by each accepted move type, and updates the probability of applying a move based on these statistics. This ensures that there is a bias towards moves which are likely to have the most effect on the objective functions at each temperature. We expect large backtracks during the early stages of an annealing run, and small backtracks during the later stages. Finally, to include a degree of intelligence in the forward move selection, we have imposed a 25% probability of selecting a decrement of 5 where appropriate.

We ran algorithms A to E ten times each for each of the five cost functions, and we ran F and G a total of ten times each (as F and G use the evaluated objective functions for completed designs, the different cost estimates do not apply). The results are tabulated in Table 1. The first row of each block lists the average number of attempted moves. The second row states the number of times the algorithm converges on its best solution, as opposed to simply visiting it in the higher temperature stages, but converging on poorer local optima. The "+" symbol indicates that in a small number of cases, the algorithm converged around the best cost, but did not actually settle on it. The third row contains the average best cost over the ten runs — the numbers in brackets list the number of runs which produced a completed list if this was less than ten. The fourth row lists the best cost obtained over the ten runs, if a completed list was obtained .

5. Discussion

First, consider the first five algorithms (those not requiring a complete generation at each move). From these results, it can be seen that in terms of approaching the optimum, original shape annealing (A), although it does find relatively short lists, is restricted by its limited backtracking strategy. In general, B, which jumps back to higher states, produces better solutions in fewer iterations. C is consistently best at finding the optimum for the first four objective functions. However, it takes at least an order of magnitude more iterations than the other algorithms, and rarely converges on its best cost. The increase in the number of iterations is largely due to the irregular cost comparison and Huang et al.'s method of calculating statistics based on the average costs of the states at a given temperature. This increased number of iterations allows the algorithm to sample more of the space, and it tends to visit the optimum cost during the early to middle stages. Its inability to converge on good solutions is also caused by the irregular comparison method. The algorithm is able to backtrack at very little cost: by comparing the costs of the states at the high level, it makes its decisions without consideration of how close the previous state was to a complete list. Finally, by comparing states at that level, decisions are taken solely in terms of the local objective functions, and thus the algorithm gets caught in the "traps" at values 60 and 30. D, which uses the same backtracking move but compares the new state with the previous state, requires significantly fewer iterations, and has better convergence. E, which extends the look ahead limit, tends to converge more frequently on better solutions in less time.

TABLE 1. Results

A	1	2	3	4	5
iterations	79495	59545	54667	46083	89052
converge	0	0 +	1 +	7 +	0
average	26.9	27.1	25.9	21.1	26 (2)
best	23	26	22	20	26

B	1	2	3	4	5
iterations	58811	51275	30704	23631	101590
converge	1 +	2	8	8	0
average	26.5	24.6	23.3	20.5	*
best	24	21	22	20	*

C	1	2	3	4	5
iterations	217000	267858	330362	366261	181698
converge	0	0	0	0	0
average	20.5	20.4	20	20	21 (2)
best	20	20	20	20	21

D	1	2	3	4	5
iterations	94447	51503	49860	31359	65747
converge	1+	1 +	4	10	0
average	24.1	23.5	22.2	20.6	21 (1)
best	23	21	21	20	21

E	1	2	3	4	5
iterations	94282	46986	48259	20633	31560
converge	0 +	2 +	4 +	9 +	8
average	23.7	22.9	21.4	20.4	20.44 (9)
best	22	22	21	20	20

F	
iterations	65662
converge	6
average	21.4
best	21

G	
iterations	9252
converge	8
average	20.7
best	20

Note also that the different cost estimates also play a significant role. The worst cost objective function (1) causes every forward move to be accepted, and thus all optimization is carried out during the backtracking moves. As the cost estimates improve, selection of forward moves becomes more discriminating. Generally, the solutions improve as the local cost estimates better reflect the final cost, except in the case of the "best cost" estimate (5), which generally causes the algorithms to converge in the local optima. The exception to this rule is E. It appears that the extended look ahead allows the algorithm to backtrack out of the local optima by jumping over the moves that trap the other variations, in both the "good cost" and "best cost" objective functions. For the first three cost functions, E is also better, but less markedly so. E's success may be attributed to the fact that it is able to jump back above local optima, and then look ahead far enough to identify better moves.

Finally, consider the performance of the two algorithms requiring complete generations. The first (F) consistently finds and converges on good solutions, in relatively few iterations, although each iteration is a longer and more complex move than in the first 5 algorithms. However, F rarely finds the optimum. G finds and converges on the optimum regularly, in fewer iterations. In turn, though, each iteration for G is on average 10 times as complex as for F, and the algorithm takes a correspondingly longer time to finish. The improved performance of G appears to be due to the sampling of final solutions compensating for the random nature of the forward generation. Suppose in the mid to late stages of a run that the move from s_i to s_{i+1} was a poor move, and that s_i appears relatively early in the move sequence. Because we are at least in the middle of a run, we will have optimized the moves below s_{i+1} to some extent. Suppose also that we now backtrack to s_i, move forward to s_{i+1}' and then complete the generation, and that the move to s_{i+1}' is better than the move to s_{i+1} in global terms. Because the list completion from s_{i+1}' is random, it is unlikely to result in a better final list than the part-optimized completion from s_{i+1}, and thus it is unlikely to be accepted, even though the move from s_i to s_{i+1}' was better. For this problem, the sampling of completions in G appears to be sufficient to direct the algorithm towards the best moves.

Although E appears to be the most successful, care should be taken in assuming that it is the best algorithm to use. The particular look ahead limit imposed here may be suited to the particular local optima found in the design space. In addition, it is probably also significant that the design space is relatively uniform away from these optima, and a simple, regular cost estimate is a good characterization of the space. With irregular spaces, it is not clear that E would perform so well. In such cases, G, although slower, may prove superior. More research is required relating the characteristics of the design space to the move set and intermediate objective functions.

6. Conclusions

Shape annealing is a robust, efficient method for generating good solutions in large design spaces. Its performance for those applications where a forward move restricts the design space can be improved by modifying its backtracking moves, incorporating a jump backtrack and limited look ahead. Annealing based on the sampling of completed designs also improves performance, although at the expense of time.

Acknowledgments

The authors would like to thank the National Science Foundation under grants DDM-9300196 and DDM-925-8090, and United Technologies for supporting this work. We also thank the members of the Computational Design Laboratory, Carnegie Mellon University, for their contributions to this work; in particular, Simon Szykman provided his implementation of the dynamic simulated annealing algorithm, and spent a lot of time helping us understand its behavior.

References

Brown, K. N., Sims Williams, J. H. and McMahon, C. A.: 1992, Grammars of features in design, *in* J. S. Gero (ed.) *Artificial Intelligence in Design '92,* Kluwer Academic Publishers, Dordrecht, pp. 287-306.

Brown, K. N., McMahon, C. A. and Sims Williams, J. H.: 1993, The role of formal grammars in the engineering design process, *Internal Report*, Department of Engineering Mathematics, University of Bristol.

Brown, K. N., McMahon, C. A. and Sims Williams, J. H.: 1994a, Features, aka The Semantics of a Formal Language of Manufacturing, *Research in Engineering Design* (to appear).

Brown, K. N., McMahon, C. A. and Sims Williams, J. H.: 1994b, Describing process plans as the formal semantics of a language of shape, *Artificial Intelligence in Engineering* (to appear).

Cagan, J., and Mitchell, W. J.: 1993, Optimally directed shape generation by shape annealing, *Environment and Planning B*, **20**, 5-12.

Cagan, J.: 1994a, A shape annealing solution to the constrained geometric knapsack problem, *Computer-Aided Design*, **28**(10), 763-769.

Cagan, J.: 1994b, Research issues in the application of design grammars, *in* J. S. Gero and E. Tyugu (eds), *Formal Design Methods for CAD*, North-Holland, Amsterdam, pp. 191-198.

Cagan, J. and Kotovsky, K.: 1994, The generation of the objective function in a simulated annealing model of problem solving, *Working Paper*, available from the authors.

Carlson, C.: 1994a, A tutorial introduction to grammatical programming, *in* J. S. Gero and E. Tyugu (eds), *Formal Design Methods for CAD,* North-Holland, Amsterdam, pp. 73-84.

Carlson, C.: 1994b, Design space description formalisms, *in* J. S. Gero and E. Tyugu (eds), *Formal Design Methods for CAD*, North-Holland, Amsterdam, pp. 121-131.

Fitzhorn, P. A.: 1989, A computational theory of design, *Preprints NSF Engineering Design Research Conference*, College of Engineering, University of Massachusetts, Amherst.

Flemming, U.: 1987, More than the sum of their parts: The grammar of Queen Anne houses, *Environment and Planning B*, **14**, 323-350.

Heisserman, J. and Woodbury, R.: 1994, Geometric design with boundary solid grammars, *in* J. S. Gero and E. Tyugu (eds), *Formal Design Methods for CAD*, North-Holland, Amsterdam, pp. 85-105.

Huang, M. D., Romeo, R. and Sangiovanni-Vincentelli, A.: 1986, An efficient general cooling schedule for simulated annealing algorithm, *ICCAD-86 IEEE International Conference on CAD, Digest of Technical Papers*, Santa Clara, CA, pp. 381-384.

Hustin, S. and Sangiovanni-Vincentelli, A.: 1987, TIM, a new standard cell placement program based on the simulated annealing algorithm, *IEEE Physical Design Workshop on Placement and Floorplanning*, Hilton Head, SC.

Kirkpatrick, S., Gelatt, C. D. Jr. and Vecchi, M. P.: 1983, Optimization by simulated annealing, *Science*, **220**(4598), 671-679.

Knuth, D.: 1968, Semantics of context-free languages, *Mathematical Systems Theory*, **2**(2), 127-145.

McMahon, C. A., Sims Williams, J. H. and Brown, K. N.: 1993, A transformation model for the integration of design computing, *International Conference on Engineering Design (ICED'93)*, The Hague.

Penjam, J.: 1990, Computational and attribute models of formal languages, *Theoretical Computer Science*, **71**, 241-264.

Reddy, G. and Cagan, J.: 1995, An improved shape annealing algorithm for truss topology generation, *ASME Journal of Mechanical Design*, **117**(2A), 315-321.

Rinderle, J. R.: 1991, Grammatical approaches to engineering design, Part II: Melding configuration and parametric design using attribute grammars, *Research in Engineering Design*, **2**(3), 137-146.

Schmidt, L. C. and Cagan, J.: 1995, Recursive annealing: A computational model for machine design, *Research in Engineering Design*, **7**, 102-125.

Stiny, G. and Mitchell, W. J.: 1978, The Palladian Grammar, *Environment and Planning B*, **5**, 5-18.

Stiny, G.: 1980, Introduction to shape and shape grammars, *Environment and Planning B*, **7**, 343-351.

Stiny, G. and March, L.: 1981, Design machines, *Environment and Planning B*, **8**, 245-255.

Stiny, G.: 1981, A note on the description of designs, *Environment and Planning B*, **8**, 257-267.

Stiny, G.: 1991, The algebras of design, *Research in Engineering Design*, **2**(3), 171-181.

Szykman, S. and Cagan, J.: 1995, A simulated annealing approach to three dimensional component packing, *ASME Journal of Mechanical Design*, **117**(2A), 308-314.

Appendix

Symbols: {<, >}

Initial list: < 100 >

X and Y are real-valued variables.

Rules:

1) $X > \;\rightarrow\; X\ \max(X - Y, 0) >$ where $X < 29.99$ or $32 \leq X < 59.99$ or $62 \leq X \leq 100$ and $Y \in \{0.01, 0.02, \ldots, 4.99, 5\}$

2) $X > \;\rightarrow\; X\ 55 >$ where $61 \leq X < 62$

3) $X > \;\rightarrow\; X\ 59.99 >$ where $60 \leq X < 61$

4) $59.99 > \;\rightarrow\; 59.99\ 59.98 >$

5) $X > \;\rightarrow\; X\ 25 >$ where $31 \leq X < 32$

6) $X > \;\rightarrow\; X\ 29.99 >$ where $30 \leq X < 31$

7) $29.99 > \;\rightarrow\; 29.99\ 29.98 >$

5

INCORPORATING HEURISTICS AND A META-ARCHITECTURE IN A GENETIC ALGORITHM FOR HARNESS DESIGN

CARLOS ZOZAYA-GOROSTIZA AND LUIS. F. ESTRADA
Instituto Tecnológico Autónomo de México, México

Abstract. This paper presents some recent results that were obtained when a basic genetic algorithm (GA) for optimizing the cost of electrical wire harnesses was modified. These modifications included the incorporation of two operators that were specific for the problem being solved: a) a *gauge propagation* operator, and b) an operator that attempts to improve a solution by randomly changing wire gauges associated with a particular device of the harness. In addition, the modified GA included the implementation of a *meta-architecture* that was useful to overcome the problem of finding a set of good input parameters for running the single-layered GA. These modifications differ from other general purpose techniques that have been suggested for improving the search in GAs. Results obtained with the modified GA for an example harness showed that the modifications were helpful for improving the effectiveness and efficiency of the basic GA.

1. Introduction

The design and optimization of the wire harnesses that compose the electrical system of a vehicle is a complex and challenging task. It involves coming up with a product that not only is easy to manufacture at a low cost, but that also satisfies a set of multiple design constraints. Some of the constraints that have to be considered when designing an automotive wire harness include physical constraints related to the physical configuration of the vehicle; thermal constraints related to the behavior of wires conducting the currents required to operate the electrical devices to which the harnesses of the vehicle are connected; voltage constraints associated with the minimum voltage that each of these devices requires to operate properly; and other constraints related to the manufacturability and maintainability of the product.

Recent studies have shown that the use of computer tools for optimizing the cost of the harness can lead to important savings in the cost of the product and to more reliable designs. Previous work with genetic algorithms (GAs) in this problem has shown that this technique can lead to lower cost solutions than those found with mathematical programming or heuristic methods (Zozaya-Gorostiza, Sudarbo and Estrada, 1994). One reason for this is that GAs search for solutions effectively regardless of the convexity of the search space (Goldberg, 1989). However, using a general purpose GA for this problem required many evaluations of possible solutions before finding low cost solutions that were comparable with those found by other methods. In addition, our experiments showed that the GA was very sensitive to the set of parameters used in its application. As a result, we decided to explore on modifying the basic GA for improving its performance in this particular problem.

In this article, we describe some modifications that were incorporated in the basic GA described in (Zozaya-Gorostiza, Sudarbo and Estrada, 1994) to improve its behavior. In particular, the incorporation of design heuristics and the development of a meta-architecture that is used to find appropriate values for the parameters of the single-layered GA were included. The modifications that incorporate design heuristics into new genetic operators differ from other techniques that have been suggested for improving the search in a general purpose GA (Booker, 1987; Bäck, 1992), as explained in section 4. The results obtained with the modified GA showed that these changes were helpful for improving the effectiveness and efficiency of the basic GA.

The article is organized as follows. First, a description of the problem being addressed is presented, with a brief discussion on the results obtained when trying to solve it by using a heuristic search program and a mathematical programming model. This presentation is followed by a description of how GA were initially applied to the harness optimization problem. Then, the modifications that were incorporated into this basic GA are described. Finally, the results that we observed in the performance of the modified GA for an example harness are discussed.

2. Harness Optimization

Harness optimization is the process that involves obtaining a set of appropriate wire gauges for a particular harness topology. It may be considered a subprocess of the broader problem of harness design. When conceptualizing a new harness, the designer creates a layout of the electrical system, decides upon its main electrical components and specifies an initial set of wire gauges and fuse sizes for the harness. Then, he or she tries to identify potential reductions in cost for a given layout of the electrical system (i.e., the harness topology) and a given set of devices to be operated in order to optimize the cost of the harness.

Harness optimization is not a simple task. On one hand, the designer tries to use as small wire gauges as possible to minimize the cost of the harness; but on the other, wire gauges have to be large enough to provide all devices with enough voltage to operate appropriately and to be able to transmit the corresponding current intensities without burning.

Figure 1 shows an example of a harness layout that may the used to illustrate the activities involved in the harness optimization process:

- Initially, the designer has the following information regarding the harness layout: length and insulation type of each of the wires (A through I), minimum current and voltage required by each device (1 through 3), connecting points for each wire and voltage intensity of the battery (+).
- Depending on the minimum current intensity required for each device to operate, each of the wires of the harness will need to transmit a particular value of current; this computation is straightforward unless the harness has switches that constrain the simultaneous operation of the devices. For example, the current transmitted by wires B and F will be at least equal to the current required by device 1 to operate; the current in wire I will be the sum of the currents in wires F and G, and so forth.
- Based on these currents, the designer identifies which is the minimum gauge that each wire needs to have in order to be able to transmit the corresponding current value. In this task, the designer is helped by a thermal model that provides information about the temperature reached by a particular wire, as a function of the current intensity being transmitted, the insulation type and thickness of the wire, the temperature of the environment surrounding the wire, and the gauge of the wire. This model also provides the designer with information about the voltage drop that will be present along the wire per unit length.
- Having identified the set of minimum gauges for the wires, the designer evaluates whether all devices have enough voltage to operate properly. If this were the case, the current harness design would be optimum for the given layout and device information. However, it is usual to find multiple devices without enough voltage to operate.
- The designer has to increase the gauges of some of the wires of the harness in order to provide all devices with the required voltages. At this point, the designer is faced with multiple choices. For example, if devices 1 and 2 of the example layout require more voltage, the designer can increase the voltage of any of the wires, except wires E and H, to provide them with more voltage. Furthermore, we can have multiple possibilities for satisfying the voltage constraints associated with these devices, and each of them has a different cost.

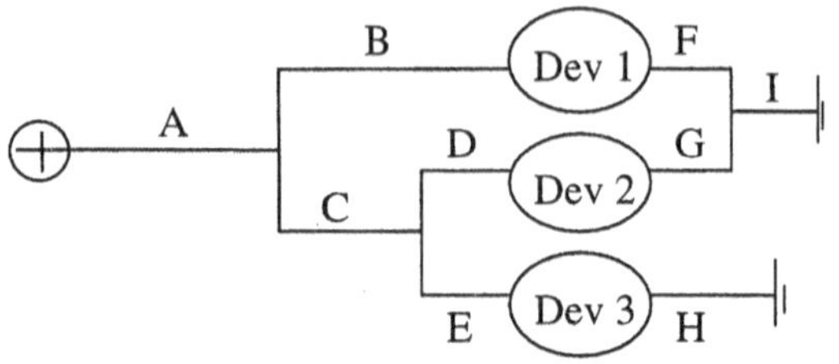

Figure 1. Example of a harness layout.

At this moment, the reader might argue that the problem can be formalized as a mathematical program, where the objective function is expressed as the minimization of the cost of the harness ($\sum$ CijXij, where Cij is the cost of wire i for gauge value j, and Xij is a binary variable that is set to 1 when wire i is assigned gauge value j), and voltage requirements are expressed as constraints of the model. In fact, this assertion is true, but only for harness layouts where all devices are connected in parallel (as in Figure 1) since all voltage constraints can be expressed as linear functions. For example, the voltage constraint associated with the first device would be expressed as follows:

$$\text{Vbatt} - \text{VdropA} - \text{VdropB} - \text{VdropF} - \text{VdropI} \geq \text{VminDev1}$$

where Vbatt is the voltage provided by the battery, VdropA through VdropI are the voltage drops observed at each of the wires that are in the path that leads from the device being considered (i.e., Device 1) to the battery and to the ground (and can be expressed as a function of the wire gauges), and VminDev1 is the minimum voltage required by device 1 to operate properly.

In more complex topologies, however, the use of a mathematical model is not straightforward. For example, consider the harness topology shown in Figure 2. For this topology, we can describe the voltage restrictions on devices 1 and 2 as linear functions; however, the voltage restriction on device 3 is not a linear function of the voltage drops observed in some of the wires of the harness. Furthermore, our research (Zozaya-Gorostiza, Sudarbo and Estrada, 1994) has shown that even if the mathematical model can be formulated, the results obtained when solving the model using powerful modeling tools such as GAMS (Brooke, Kendrick and Meeraus, 1987) are not optimal.

An initial approach in trying to solve the harness optimization problem was the development of OPTAR (Zozaya-Gorostiza, 1991). OPTAR implemented a *hill-climbing* heuristic search procedure in which the gauge of a *selected* wire was increased to the next allowable gauge value for each iteration of the algorithm until all devices had enough voltage to operate.

Wires were selected using a heuristic formula that included the number of devices that were affected by each wire, the marginal cost incurred when increasing the corresponding gauge, and the additional voltage supplied when doing this gauge change. The results obtained with OPTAR received encouraging comments from harness designers of the Packard Electric Division of General Motors, because at that moment the only tools available to support the harness design process were some simulation packages and empirical models (O'Keefe, 1989; Styer and Burns, 1990).

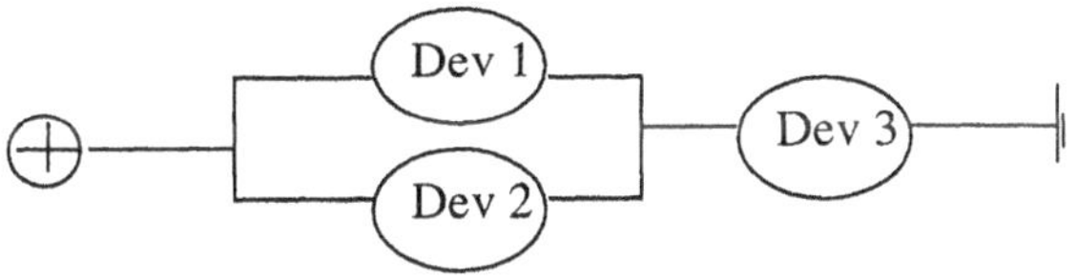

Figure 2. Example of three devices that are not connected in parallel.

The next attempt was to use genetic algorithms for improving the results obtained by OPTAR. This technique had shown good results when applied to other design and optimization problems, and have the power to search in complex solution spaces regardless of their shape. In the next section we summarize how we initially used this technique for solving the harness optimization problem.

3. Basic Genetic Algorithm

Genetic algorithms (GAs) are search methods based on the mechanics of natural selection and genetics (Holland, 1975). They employ string structures (called *chromosomes*) to represent sets of solution variables and a *fitness function* to evaluate these sets. New solutions are obtained through a combination of the material included in these strings by means of different genetic *operators*. A simple genetic algorithm uses various kinds of random operators: a *selection operator* identifies those chromosomes that may be used to generate new chromosomes, a *crossover operator* creates two children chromosomes by randomly exchanging portions of the parent chromosomes, and a *mutation operator* randomly modifies parts of the strings. In addition, a GA has a replacement operator that inserts and replaces chromosomes of a certain population to create a new one. Starting with an *initial population*, the GA proceeds iteratively until a stopping criteria is achieved. Each iteration in which a new population is obtained is known as a *generation* (see Figure 3).

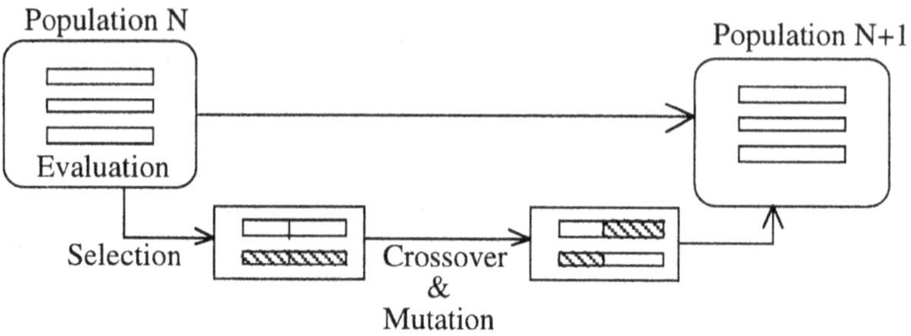

Figure 3. Operation of a basic GA.

In our initial application of a GA to the harness optimization problem, we used binary strings to represent the set of wire gauges associated with a particular harness design. Each wire gauge was represented by 4 bits, and the chromosome was formed by concatenating these segments for all the wires in the harness. The order in which these 4 bit segments are concatenated is completely independent from the harness topology (see Figure 4).

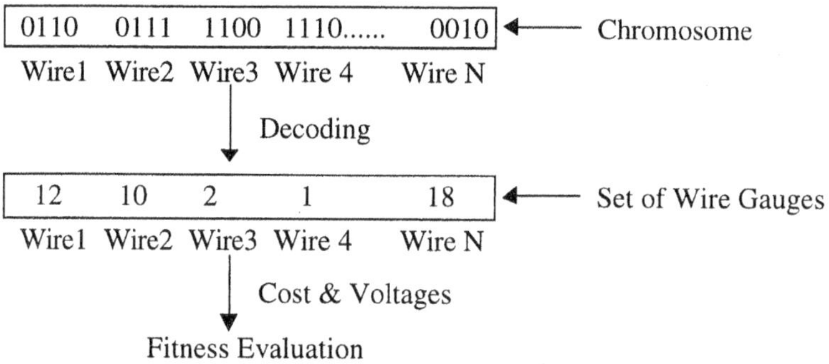

Figure 4. Harness representation and decodification.

The initial population was randomly generated by taking into consideration all possible gauge values a wire might have (22, 20, 18, 16, 14, 12, 10, 8, 6, 4, 2, 1 and 0), and the evaluation function was responsible for a) decoding the chromosomes into a set of gauge values, b) computing the fitness of the chromosome by considering the cost of the solution being represented, the violation to the voltage constraints of the devices and the violation of the thermal constraints on each wire. In the basic GA, decoding is performed by linearly mapping the 16 possible values obtained by the 4-bit segment to the 13 possible gauge values; for example, if the 4-bit segment consists is 1110, the associated gauge value would be the 12th element (i.e., round(13*(14+1)/16)= 12) of the set of possible gauge values, which is equal to value 1. The selection, crossover, mutation and replacement operators were used to generate the next population of solutions, and the procedure continued until a predefined value of generations was achieved.

Figure 5 shows an example of how the crossover operator is applied to a pair of parent chromosomes. A random number that represents the position of the string that will be used to exchange the genetic material of the two parents, called the locus for crossing, is generated. The children chromosomes are generated by exchanging left and right portions of the two chromosomes with respect to the locus position. It is interesting to note that even though the only wire whose gauge changed in the children chromosomes was wire 3, the set of wire gauges in the children chromosome represent different configurations for the harness.

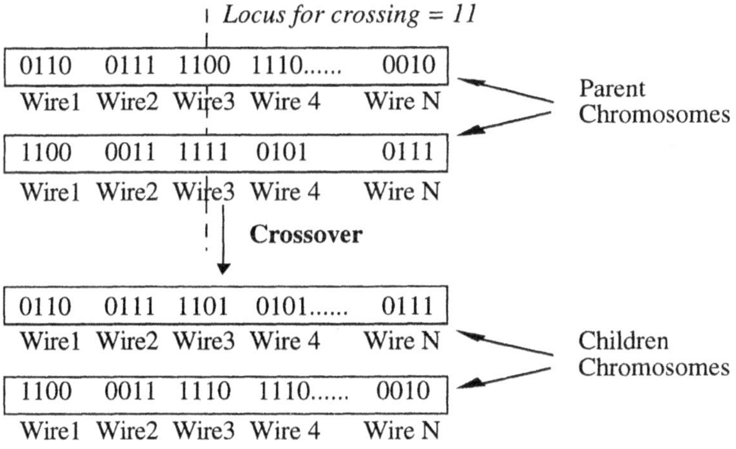

Figure 5. Crossover operator.

4. Modified Genetic Algorithm

The results obtained with the basic GA described in the previous section were stimulating. In fact, we were able to obtain better solutions than those obtained with the GAMS model. However, we found that the basic GA required a large number of evaluations to obtain designs that were comparable in cost and performance than those obtained by the other techniques (i.e., mathematical programming and heuristic search) described earlier.

In order to improve the efficiency and effectiveness of the GA, we decided to incorporate various heuristics associated with the design of electrical wire harnesses. These heuristics include:

- including chromosomes with minimal gauges in the initial population;
- using an operator that modifies the chromosomes according to a design heuristic;

- mapping the gauge values obtained when decoding the chromosomes to values greater than or equal to the minimal gauges; and
- using an operator that inserts new chromosomes into the population.

In addition, we found that the basic GA was very sensitive to its input parameters. After various attempts to obtain a good and robust set of parameters, we implemented a meta-architecture that could provide us with appropriate input parameters. In the discussion that follows we describe how these modifications to the GA were implemented.

4.1. INITIAL POPULATION

In the basic GA, the initial population was generated randomly by allowing each wire to take any possible gauge value. In this population, each solution (i.e., chromosome) might have gauge values that violate the thermal restrictions associated with the type of insulation of the wire.

In the modified GA, the initial population is obtained by using the minimal gauges that are thermally feasible for each wire. Some chromosomes with these values are introduced in the population, and the rest of the population is generated by using the crossover and mutation operations on these chromosomes. These genetic operators are applied with high probability values (0.8 and 0.01 respectively) to ensure that there is enough diversity in the initial population. After these operators are applied, resulting gauge values are mapped to the set of thermally feasible gauge values to ensure that there are no violations in the thermal constraints of the design problem.

4.2. GAUGE PROPAGATION

A simple design heuristic in the case of electrical wire harnesses requires that the gauge of a cable that splits into two or more cables is greater than or equal to the gauges of these cables. Conversely, if two or more cables join into another cable, the former cables have to have a gauge value that is less than or equal to the gage of the latter cable. Figure 6 illustrates this heuristic: the gauge value of cable A has to be greater than or equal to the gauge values of cables B and C, and cables X and Y have to have gauge values that are less than or equal to the gauge value of cable Z.

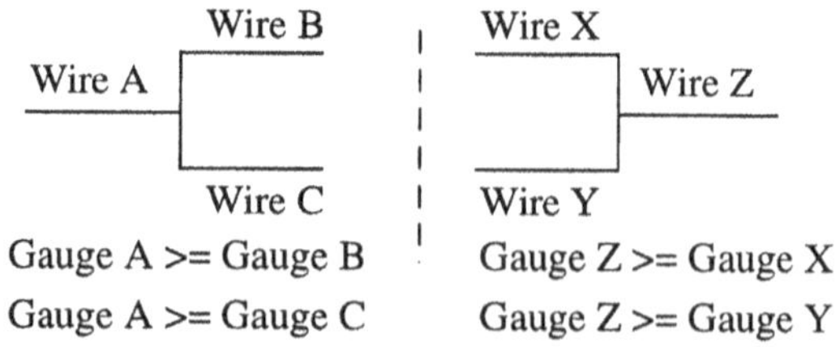

Figure 6. Gauge heuristic.

In a GA, this heuristic can be violated when the crossover or mutation operators are applied to generate new chromosomes in the population. In the modified GA, we use a *gauge propagation* operator that adjust gauge values to comply with this heuristic. This operator allows the GA to reduce wide gauge values in wires that, because of their location in the topology of the harness, may have thinner gauge values.

The application of the *gauge propagation* operator is not deterministic. Even with the heuristic it is impossible to know if the propagation has to be "upwards" (i.e., towards the devices or the harness) or "downwards" (i.e., towards the battery or the ground). Considering the examples shown in Figure 6, and assuming that in a particular chromosome the gauge of wire A is less than or equal to the gauge of wire B, we have two manners of complying with the gauge heuristic:

- to increase the value of wire A, which takes us to a harness with higher cost but more voltage to those devices that are affected wire A;
- to decrease the value of wire B, which takes us to a harness with lower cost but less voltage to those devices affected by wire B.

In the GA, the voltage available to each device of the harness is computed by the fitness function of the algorithm. Therefore, it is impossible to know which decision might be more appropriate. Furthermore, we did not want to create a deterministic operator that would constrain the search for new solutions. As a result, the *gauge propagation operator* is applied either "upwards" or "downwards" randomly, and the wire that is used as the starting point in this operation is also selected in a random manner.

Another possibility that could have been explored to comply with the design heuristic would have been to modify the crossover and mutation operators directly. This modification would have implied to alter these operators so that the chromosomes that are being generated do not violate the heuristic. For example, in the crossover operator, we could have modified the manner how the locus for crossover is selected, or we could have implemented a loop to apply this operator as many times as needed until a new individual that complies with the heuristic is obtained. It is also possible to include the heuristic as part of the evaluation function so that chromosomes that comply with it have higher fitness values.

In our GA we decided not to alter the basic GA operators directly. The *gauge propagation operator* is applied randomly using a new input parameter called the *propagation probability.* We generate a random number for each chromosome in the population, and if it is less than or equal to the propagation probability, the propagation process is applied. In this case, a new random number is generated to select the wire that will be used as the starting point of the propagation and another number to decide whether the propagation is "upwards" or "downwards". Then, we search in the chosen selection for wires that violate the design heuristics and alter their gauge values to comply with it. The random selection of the wire chosen as the starting point allows the GA to incorporate new genetic material into the population and to obtain different modified individuals even if the operator is applied in identical chromosomes. In addition, the compliance with the design heuristic was also included as part of the evaluation function in order to penalize those individuals that violate this heuristic. Figure 7 shows an example of the application of the gauge propagation operator for a sample harness topology.

Harness topology

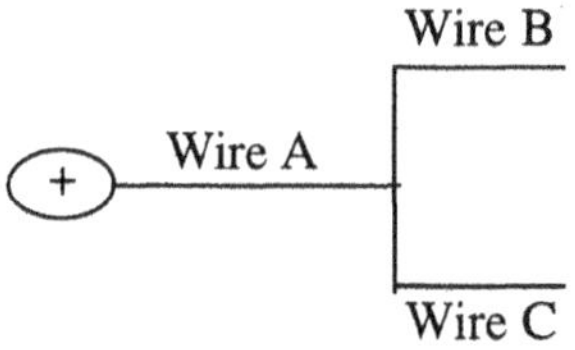

Original Gauge Values		
Wire A	Wire B	Wire C
16	12	18

Type of Propagation	From Wire	Gauge Values after Propagation		
		Wire A	Wire B	Wire C
Upwards	A	16	16	18
Downwards	B	12	12	18
Downwards	C	18	12	18

Figure 7. Gauge propagation.

Gauge propagation can be undertaken easily by representing the wire harness by double linked lists. This representation allows the operator to navigate upwards or downwards in the structure of the harness from a particular wire. Recursive functions are used to implement this navigation. Two data structures are used to represent the topology of the harness. The primary structure stores the identifier of the cable and two pointers, one that points to the "father" wire of the cable, and one to the first of its "sons". The second structure is used to identify all cables that are at the same level of ramification.

4.3. GAUGE MAPPING

Another type of heuristic that was implemented consisted of mapping the values that are obtained from the execution of the crossover and mutation operators. Since we use 4 bits to represent each wire, the set of 16 possible values that may be obtained with these four bits have to be mapped to the set of possible gauge values. In a simple GA, each wire might have any of the

13 gauge values described earlier. However, some of these values might violate the thermal constraints associated with its type of insulation. For example, if a wire transmits a high current value, a gauge value of 22 or 20 might cause its insulation to melt. Therefore, we represent the minimal gauge values that were obtained using the thermal model RADWIRE, and map the 16 possible 4-bit values to those wire gauges that are wider than these minimum values. This modification of the simple GA ensures that all the individuals of the population comply with the thermal constraints of the model. In other words, this mapping allows us to exclude the satisfaction of the thermal constraints from the fitness function of the algorithm.

Gauge mapping is implemented in a straightforward manner. The value associated with the 4 bits of a wire is divided by 16 and multiplied by a number that represents the size of the set of possible gauge values for this wire. If all gauge values are permitted (i.e., the cable transmits a low current intensity), this number is 13. However, for wires with higher current intensities, this number might range from 13 to 1. The result is used as an index for finding the corresponding wire gauge.

4.4. INSERTION OF IMPROVED CHROMOSOMES

When a GA is being executed, the fitness of the solutions being generated improves through the iterations of the algorithm. However, the convergence of the algorithm is slower as the algorithm finds better solutions to the problem. In the case of our GA, the basic algorithm converged rapidly during the first generations (100 to 200) but it found difficult to improve this solutions after these initial iterations.

In order to deal with this problem, several authors have discussed mechanisms to ensure the diversity of the population (Goldberg, 1989). However, our experiments showed that diversity was not the cause of the problem; what we needed was a mechanism that would fine-tune the solutions being proposed in order to improve the effectiveness and efficiency of the algorithm.

We modified the basic GA to include an operator that a) proposes cheaper solutions for a given chromosome by trying to reduce the gauges of some of the wires without affecting the performance of the devices of the harness, or b) attempts to satisfy the required voltage for particular devices. This operator takes the best solution found in a given population (i.e., the one that has the higher value of the fitness function), picks a particular device of the harness, and explores whether it can reduce or increase the gauges of the wires that are in the path that leads from this device to the battery, or from this device to the ground, in order to reduce the cost of the harness without violating the constraint associated with the minimum voltage of this device, or provide it with enough voltage to operate.

The operator proposes solutions using a procedure that resembles a *hill-climbing* type of search. Since this technique can lead to local optima, we decided to use it carefully in order not to affect the benefits of using a GA. The operator is executed only after the GA has been run through a given number of iterations, in order to ensure that it has already explored a good portion of the search space.

The operator that proposes an improved solution for a given chromosome, works as follows:

1. We give the GA as inputs several variables that indicate whether this operator should or should not be applied, the starting generation for using this operator, and a constant that represents the gap between generations in which the operator should be applied.
2. When the modified GA gets to the point at which the operator may be applied, the algorithm picks the best solution found in the preceding population.
3. The GA identifies those wires that are associated with the first device of the harness (i.e., those whose gauge values affect the available voltage in this device) and selects one of them randomly.
4. The gauge of this particular wire is decreased to the next allowable lower value of those thermally feasible. As a result, the available voltage for the device and the total cost of the harness are reduced. The algorithm then checks whether the selected device has still enough voltage to operate. If this is the case, the algorithm goes to step 8; otherwise, the gauge of the wire is restored to its original value and the algorithm goes to step 5.
5. If the device being analyzed does not have enough voltage to operate, the gauge of the selected wire is increased to the next allowable higher value. As a result, the available voltage for the device is also increased. The algorithm then checks whether the selected device has enough voltage to operate. If this is the case, the algorithm goes to step 8; otherwise, the gauge is restored to its original value and the algorithm goes to step 6.
6. In the case in which no improvements could be made in the cost of the harness (by reducing the gauge of the selected wire in step 4) or in the satisfaction of the voltage constraint of the device (by increasing the gauge of the selected wire in step 5), the algorithm picks randomly another cable of those associated with the device. If all the wires identified in step 3, or if five of these wires have already been selected, the algorithm stops; otherwise, the algorithm returns to step 4 with the newly selected wire.
7. If none of the gauge changes proposed was useful for satisfying the voltage in the device or for reducing the cost of the harness, the algorithm selects randomly one of the wires identified in step 3 and applies the *gauge propagation* operator. This is performed in order to force the insertion of new genetic material into the population.

8. The new combination of wire gauges is introduced in the best individual of the preceding generation, and the modified chromosome is included into the new population.
9. The algorithm repeats steps 4 through 8 for the remaining devices of the harness, so that one new individual is introduced into the new population for each device.

4.5. META-ARCHITECTURE

One of the problems that we found when using GA to optimize wire harness costs was the selection of appropriate input parameters for the GA. Even with the incorporation of the mechanisms described in the previous section, the performance of the GA was very sensitive to the set of input parameters being used. Initially, we attempted to find a set of good input parameters using a manual trial-and-error procedure. However, the results obtained with this method were not encouraging. Therefore, we decided to explore other alternatives and an innovative idea was to use a *meta-population* in which the *meta-chromosomes* represented various combinations of input parameters for running the modified GA. This allowed us to use the same genetic operators in order to find the set of inputs we were looking for.

In the *meta-population*, each individual (i.e., meta-individual represented as a meta-chromosome) stores information of the following six input parameters for the modified GA (see Figure 8): three associated with the probabilities in which the GA operators will be applied (crossover, mutation and gauge propagation), and three associated with the relative weight for each term of the fitness function (satisfaction of voltage constraints, cost of the harness, and satisfaction of the design heuristic). Again, we used 4 bits for representing the values of each of these parameters, and we encapsulated these bits using two unsigned integers in a similar manner how we represented the wires in the regular population.

Unsigned integer 1				Unsigned integer 2	
Crossover Probability	Mutation Probability	Propagation Probability	Voltage Weight	Cost Weight	Propagation Weight
1011	0011	1101	0010	1101	0011
Probabilities for applying the different Operators			Relative weights for each term of the Fitness Function		

Figure 8. Representation of a meta-chromosome.

Mapping of values was done by considering that the crossover and propagation probabilities could have values that range from 0 to 1 in increments of 0.1, and that mutation probabilities could have values ranging from 0 to 0.01 in multiples of 0.001. Similarly, the allowable values for the relative weights of each of the parameters in the fitness function was done by taking a range from 0 to 100% in multiples of 10% for each of the weights. However, since we want to ensure that the sum of the relative weights is always set to 100%, the GA adjusts these weights by increasing or decreasing them until this sum is satisfied. If the sum is greater than 100%, the algorithm reduces the relative weights associated with the satisfaction of the design heuristic, the voltage constraints or the cost, in this order, until the sum is set to 100%. If the sum is less than 100%, the algorithm increases the weight values in the opposite order until the sum is satisfied.

Figure 9 shows the operation of the Meta-GA (i.e., the modified GA with a meta-architecture). For each meta-individual (1 through M), the lower level GA is run independently through N generations with the set of input parameters stored in the meta-individual. At the end of each meta-generation, each meta-individual is associated with the best solution found in the corresponding population N of the lower level GA, and the GA will proceed to combine the genetic material of these meta-individuals to come up with new sets of parameters (i.e., new meta-individuals) for running the modified GA.

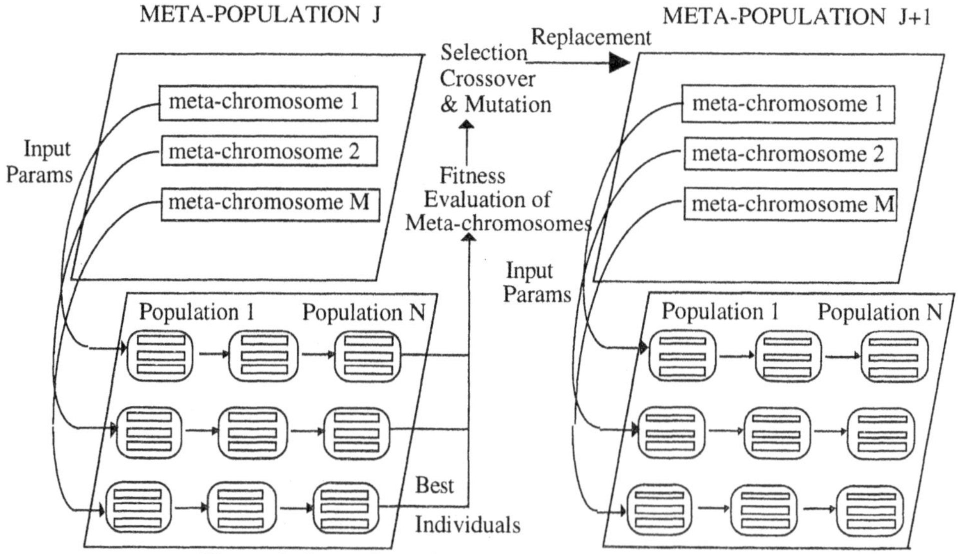

Figure 9. Operation of the meta-genetic algorithm.

In order to provide a GA with a self-adaptation mechanism, other authors such as Bäck (1992) have incorporated additional bits into the chromosomes of a the GA for representing varying mutation probabilities. With this representation, each individual of the population may have associated a different set of mutation probabilities that control the execution of the mutation operator. Our solution differs from this idea for three main reasons: a) we provide the algorithm with the possibility of changing not only the values of mutation probabilities, but also the five other parameters of the GA described in Figure 8 (i.e., crossover and propagation probabilities, and the three relative weights of the fitness function); b) the adaptation of parameters in our modified GA is performed by combining the genetic material in the meta-chromosomes of the meta-population, and not by combining the genetic material in the chromosomes of the lower level population; c) for each meta-individual, the parameters encoded in its representation are taken as constant input parameters through all the N populations of the lower level GA associated with it.

At this point it is important to consider how to rank the meta-individuals in order to select which of them are more appropriate for being used for the application of crossover and mutation operators in the meta-population. Initially, one could consider using the fitness of the best individual in the lower level population as the basis for comparison. However, the problem with this alternative is that this fitness depends on the relative weights assigned to the voltage, cost and design heuristic, which are parameters encoded in the meta-individual. Therefore, an individual which has a good fitness value in one of the populations might have a bad fitness value in another. In other words, the fitness value associated with the best individuals cannot be used directly to rank the meta-individuals.

In the literature, we found no straightforward recommendation for solving this problem; in fact, we did not find a reference where a meta-architecture is implemented in a GA to find a good set of parameters for a lower level GA, since other solutions deal only with single-layered GAs (Booker, 1987; Bäck, 1992). As a result, we thought of two alternatives:

- To evaluate the best individuals associated with the meta-chromosomes using a common set of weights (i.e., a meta-fitness function); or
- To evaluate the best individuals associated with the meta-chromosomes by considering only the cost of the harness, and penalizing the evaluation by taking into account how many devices do not satisfy the voltage constraints.

The problem with using the first alternative was that we need to come up with a set of parameters for the meta-fitness function. For the second alternative we used the following formula, which is expressed as a function of terms that do not depend on a set of predefined weights:

fitness = *(minimum cost/cost of the best individual) * (no. of satisfied devices/no. of devices)*

where:

- *minimum cost* is the cost of a harness with the minimum wire gauges that satisfy the thermal constraints.
- *cost of the best individual* is the cost of the individual with highest fitness in the last generation of the lower level GA obtained using the parameters encoded in the meta-chromosome.
- *no. of satisfied devices* is the number of devices that have enough voltage to operate appropriately in the individual with highest fitness in the last generation of the lower level GA obtained using the parameters encoded in the meta-chromosome.
- *no. of devices* is the total number of devices in the harness being designed.

Once each meta-individual has an associated fitness value, the genetic material of the meta-individuals is exchanged using regular crossover and mutation operators. These operators are used to create new meta-individuals which are included in the new meta-population. As in the lower level GA, we include the best meta-individual from the previous generation into this new population to ensure a monotonic behavior of the GA.

5. Results

To measure the performance of our GA we used the same example harness that we had used in our previous research with heuristic search (Greiff and Zozaya-Gorostiza, 1989), mathematical programming and a basic GA (Zozaya-Gorostiza, Sudarbo and Estrada, 1994); this harness has 26 wires and 7 devices. The purpose of using the same harness was to be able to isolate the benefits obtained when using the meta-architecture and the new operators described in the previous section. As mentioned earlier, this harness has all its devices connected in parallel, and therefore it can be modeled using a simple mathematical program.

In the following graphs, we present the average results obtained by doing ten runs of the GA for a given set of parameters. Each point corresponds to the best individual found on multiples of ten generations. All the solutions satisfy the thermal restrictions on the wires, since we always decoded the chromosomes by mapping to allowable gauge values that do not violate these constraints. Also, since we are interested only in solutions that provide enough voltage to the devices, we only plot those solutions that satisfy this second type of constraints. Therefore, if the algorithm was run for a total of 400 generations, the graph might have less than 40 points.

Graphs are ordered to illustrate the benefits obtained when a new operator was included in the basic GA. The performance of the GA when all the modifications as well as the meta-architecture were incorporated is shown in Figure 13. The other graphs show the performance of the GA without using the meta-architecture, for a given set of input parameters.

5.1. RESULTS OF THE BASIC GA

Figure 10 shows the performance of the basic algorithm through 400 generations for different values of the crossover and mutation probabilities. The algorithm performs better for high values of the crossover probability and for low values in the mutation probability.

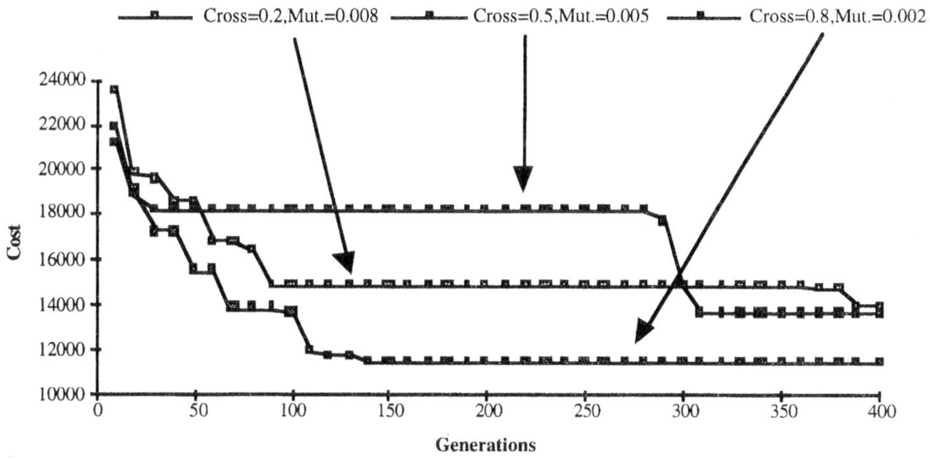

Figure 10. Performance for different values of crossover and mutation probabilities (Cost weight: 0.4, Voltage weight: 0.6, Population size: 30)

5.2. EFFECTS OF THE GAUGE PROPAGATION OPERATOR

As mentioned earlier, gauge propagation has the objective of making more effective the performance of the GA by obtaining solutions that comply with the design heuristic described in section 4.2. This operator is randomly applied to increase or decrease wire gauges in order to eliminate those cases in which a wire that transmits a high current intensity has a lower gauge value than a wire that transmit less electrical current.

Figure 11 shows the results obtained with the gauge propagation operator for different values of the probability associated with its application, and for the same set of crossover and mutation probabilities. The first graph corresponds to the case in which the operator is never applied, and is included here for comparison purposes. As in the previous graph, only solutions that provide enough voltage for the seven devices of the harness are plotted.

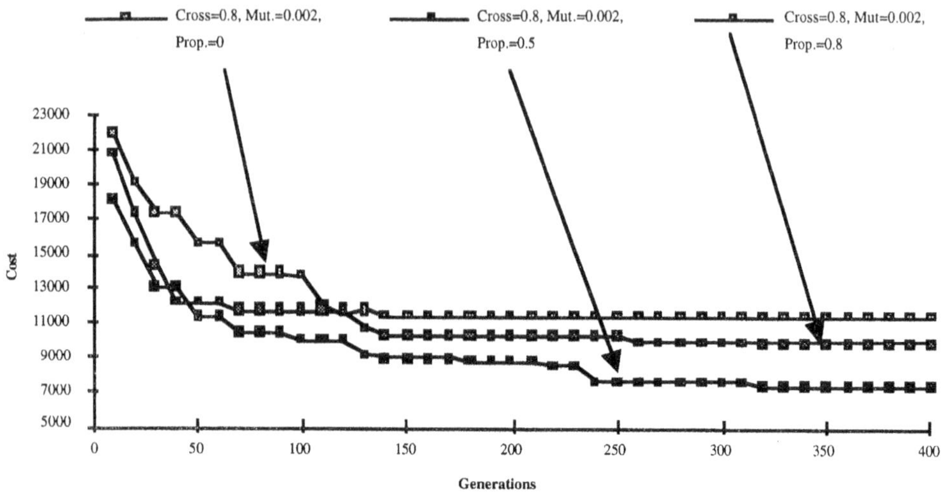

Figure 11. Performance of the GA with the Gauge Propagation Operator (Cost weight: 0.4, Voltage weight: 0.6, Population size: 30)

It is interesting to note that the best result was obtained when the probability of applying the gauge propagation operator was 0.5 and not 0.8. This fact might indicate that too much manipulation of the chromosomes might affect the power of the GA to search the solution space. By introducing the new operator we are trying to converge more rapidly to good solutions; however, the theory of GA is based on letting the traditional operators (i.e., crossover and mutation) to act freely in the population. Nevertheless, the results were in both cases better than those obtained when the operator was not applied.

5.3. EFFECTS OF THE INSERTION OF IMPROVED CHROMOSOMES

The insertion of improved chromosomes has the objective of modifying a particular solution to reduce its cost or to satisfy the voltage constraints associated with the devices of the harness. Once an improved solution has been obtained from the best individual in a given population, the chromosome that represents this solution is inserted in the succeeding population. The combination of its genetic material with that of other chromosomes is made by the crossover operator of the basic GA.

Figure 12 shows the performance of the GA with and without the application of the operator that inserts improved chromosomes into the population. The first 400 generations were run without applying this operator; starting with generation 400, every ten generations the operator was applied, and the algorithm was run until generation 700. The graph shows only those solutions that satisfied the voltage constraints for all the devices of the harness.

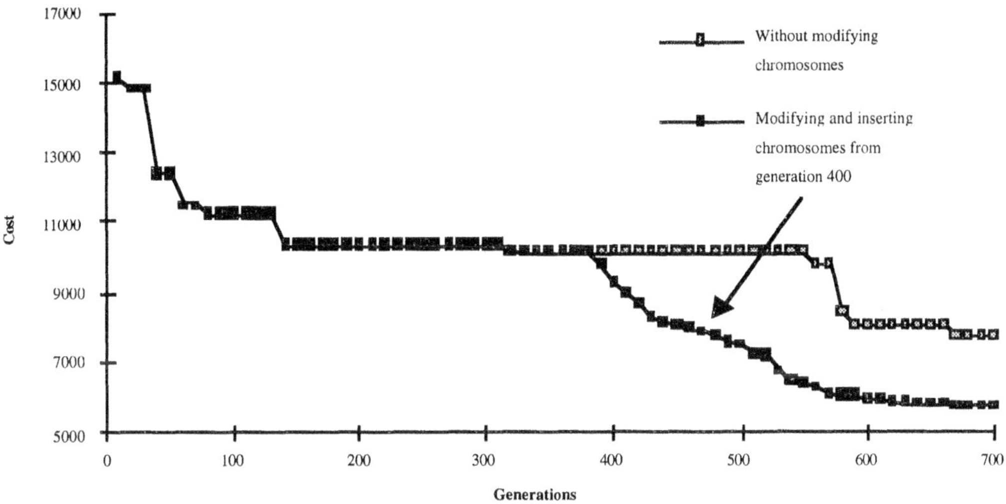

Figure 12. Performance of the GA when the operator that modifies the best individual in the preceding population is applied starting at generation 400. (Crossover prob: 0.8, Mutation prob: 0.002, Gauge propagation prob: 0.5, Cost weight: 0.4, Voltage weight: 0.5, Heuristic weight: 0.1, Population size: 30)

5.4. EFFECTS OF USING THE META-ARCHITECTURE

Figure 13 shows the performance of the GA after iterating through 40 meta-generations with 6 meta-individuals in the meta-population. In this case, we plotted the cost of the best individual obtained in each meta-population. To generate this graph, the lower level GA was run through 700 generations, using the operator that modifies the best individual in the preceding population starting at generation 400. Therefore, the set of solutions that had to be tested was 700 for each of the 6 meta-individuals, giving a total of 700*6*40= 168,000 generations; where each set of 700 generations was run with a particular set of the input parameters discussed in section 4.5. Since each generation has 30 individuals, the number of solutions that were evaluated is $5.04 * 10^6$ which is still a very small fraction of the solution space (13 possible gauge values for each of the 26 wires of the example harness, gives a total of $9.17 * 10^{28}$ possible solutions).

In each series of the graph, the manner how the GA evaluated the fitness of the meta-individuals was different. In the first series, each meta-individual is evaluated by applying a common fitness function to the best individual obtained after the modified GA was run through 700 generations using the parameters encoded in its chromosome. In the second series, we applied the same kind of evaluation, but we preserved in the meta-population the best meta-individual of the previous generation (i.e., elitism was incorporated). Finally, in the third series the meta-individuals were evaluated by using the formula described in section 4.5. In this series, elitism was also applied.

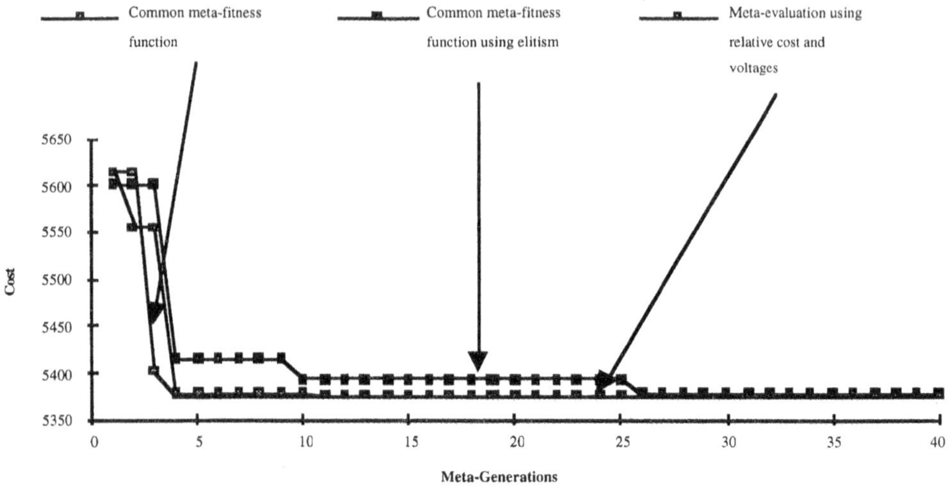

Figure 13. Performance of the GA when the Meta-Architecture was implemented. (Crossover prob: 0.8, Mutation prob: 0.002, Gauge propagation prob: 0.5, Cost weight: 0.4, Voltage weight: 0.5, Heuristic weight: 0.1, Population size: 30; Generations for each meta-individual: 700; Meta-population size: 6)

The graph shows that incorporation of elitism did not lead to significantly better results in the performance of the meta-GA. This can be explained by considering that the initial population of a meta-individual is generated by using the best individual of the populations associated with the meta-individuals whose genetic material was combined to create the new meta-individual; therefore, these individuals are likely to be preserved in the new population unless there were a drastic change in the input parameters used to run the lower level GA. As a consequence, the effect is the same as if the original meta-individuals had remained in the meta-population.

The graph also shows that the results obtained when using a common fitness function for the meta-individuals were similar to those obtained when we apply the formula that evaluates the meta-individuals using relative factors with respect to the solution with minimum gauges. This can be explained for those cases in which the common fitness function has similar relative weights for its cost and voltage weights.

6. Conclusions

This paper presents some recent results that were obtained when a basic genetic algorithm (GA) for optimizing the cost of electrical wire harnesses was modified. These modifications included the incorporation of two operators that were specific for the problem being solved: a) a *gauge propagation* operator, and b) an operator that attempts to improve a possible solution by randomly changing wire gauges associated with a particular device of the harness. In addition, the modified GA included the implementation of a *meta-architecture* that was useful to overcome the problem of finding a set of good input parameters for running the single-layered GA.

The results obtained when trying to optimize the design of an example harness show that the incorporation of domain heuristics, as well as the use of a meta-architecture in a GA, can lead to significant improvements in the performance of the GA.

These modifications could be incorporated in other applications of GAs for design activities that present similarities with the harness optimization problem. For example, the design of hydraulic networks with a given topology could be analogous to this problem by replacing voltages with pressures and current intensities with flows. However, the techniques here suggested, in particular the use of a meta-architecture, might also be applicable to generic implementations of genetic algorithms.

Further research and experimentation with other applications of GAs could help to answer questions that remain open with respect to the manner how meta-chromosomes can be evaluated in a two-layered GA. The two alternatives that were implemented in this work constitute only some of the possibilities that could be tested in the future. In addition, the convenience of implementing new operators similar to those used for the gauge propagation and chromosome improvement processes would have to be evaluated when using GAs in other types of design problems.

References

Bäck, T.: 1992, Self adaptation in genetic algorithms, *in* F. J. Varela and P. Bourgine (eds), *Toward a Practice of Autonomous Systems,* MIT Press, Cambridge, Mass., pp. 263-271.

Brooke, A., Kendrick, D. and Meeraus, A.: 1987, *GAMS,* Scientific Press, Redwood City, California.

Booker, L.: 1987, Improving search in genetic algorithms, *in* L. Davis (ed.), *Genetic Algorithms and Simulated Annealing*, Morgan Kaufmann, Los Altos, California, pp. 61-73.

Goldberg, D. E.: 1989, *Genetic Algorithms in Search, Optimization and Machine Learning*, Addison-Wesley, Reading, Mass.

Greiff, W. and Zozaya-Gorostiza, C.: 1989, OPTAR: A system for the optimization of automotive electrical wire harnesses, *Technical Report,* Condumex Harness Division, Mexico D.F.

Holland, J.: 1975, *Adaptation in Natural and Artificial Systems*, University of Michigan Press, Ann Arbor, Michigan.

O'Keefe, T.: 1989, Basic Harness Circuit Design, *Internal Report*, Packard Electric Division, Warren, Ohio.

Styer, J. P. and Burns, C. D.: 1990, Electrical system simulator and optimizer, *Internal Report,* Packard Electric Division, Warren, Ohio.

Zozaya-Gorostiza, C.: 1991, Use of AI-based tools in a Mexican automotive part supplier, *in* F. Cantú-Ortiz (ed.), *Expert Systems Applications in Mexico*, Pergammon Press, Oxford, pp. 124-144.

Zozaya-Gorostiza, C., Sudarbo, H. and Estrada, L. F.: 1994, Use of genetic algorithms to optimize the cost of automotive wire harnesses, *in* R. Baeza-Yates (ed.) *Computer Science 2: Research and Application*, Plenum Press, New York, pp. 103-115.

6

GENERATION AND SEARCH METHODS IN DESIGN: DISCUSSION

KEN BROWN
Carnegie Mellon University, USA

1. Introduction

Design *generation* is the process of formulating design solutions within a design space. An early view of design as a *search* process was discussed by Simon (1969), in which knowledge can be modeled as a set of operators, and design problems modeled as goals; design is then the process of searching for sequences of operators which produce solutions to those goals. Depending on the nature of the operators, this process may involve the refinement and transformation of incomplete designs until a complete solution is obtained, or a traversal through a space of possible solutions occurs, or a combination of both. The process may be deterministic, in which the space is searched systematically, according to fixed procedures, or stochastic, in which probabilistic methods are used to sample the space. Design as search has been criticised as being too restrictive to capture the real nature of design. Smithers *et al.* (1994) argue that a large part of design is concerned with formulating the problem and discovering the relationship between the design goals and design knowledge, and thus that design is better captured by the term *exploration*. The session summarized here considered methods of formally describing and searching spaces in order to generate design solutions, and discussed the adequacy of search as a characterization of design.

Two papers were presented in this session. Brown and Cagan (1996) present some results on the use of shape annealing for searching large state spaces. Moves within a space are selected probabilistically, where a move may include backtracking to an arbitrary height. The result of each move is evaluated, and is then either accepted or rejected with a probability derived from the evaluation. The algorithm converges to good solutions. The paper presents evidence that both the evaluation function and the details of the backtracking moves are important to the algorithm's success. Zozaya-Gorostiza and Estrada (1996) also investigate searching large spaces, although this time using genetic algorithms, which apply moves probabilistically to populations of designs. Again designs are evaluated, and multiple designs are selected to remain in the population. The paper demonstrates that both domain-specific heuristics and search in the meta space are important factors in the algorithm's performance.

These two papers have many features in common. Both considered large spaces, for which systematic search methods were assumed to be unsuitable; the ability to evaluate designs or states in the space was assumed; moves within the space were selected using random or probabilistic methods; and all three relied on statistical convergence to settle on good designs. Brown and Cagan's method considered only one candidate at a time, but ensured diverse paths through the space were considered by only gradually introducing stability as the search progressed. Zozaya and Estrada's method ensured diversity through maintaining multiple options throughout the search.

The discussion which followed was structured around three questions arising from these presentations:

1. Is the notion of random generation followed by evaluation and selection an adequate characterization of design?
2. Are there other methods of searching large design spaces?
3. Is modifying existing algorithms to suit particular problems a better approach than recoding the problem descriptions?

The text that follows does not strictly adhere to this structure, but reflects the flow of the discussion.

2. Design as Random Generation and Selection

The papers in the session presented methods in which design consisted of a series of more or less random moves to produce new states, interleaved with evaluation and selection. Is this a reasonable view of design? Firstly, it seems far from human design practice, in which steps are chosen with specific goals in mind. This raised the question of whether design is a human activity, or a more general phenomenon of which human design happens to be the best example. Secondly, assuming this approach is design, why should it be considered? One of the main reasons is that it is relatively easy to implement in computer systems, as it requires very little domain-specific design knowledge. There is no need to partition the search spaces, or to encode routines for deciding which options should be pursued in which circumstances. Once the space is described in terms of primitive elements and legal operators, generation is knowledge-free. All the knowledge is embedded in the evaluation function (and thus is only concerned with the designed artifacts and not the structure of the space used to create them).

This approach views design as more of a mathematical optimization process than a human-centered artistic endeavour.

The approach above is dependent on the assumption that the knowledge for evaluating designs is available. In many design domains, and in architecture in particular, this is not necessarily true. A distinction has to be made between book-keeping evaluation, in which a design is assessed against specific well-defined criteria—for example, conformance to building standards, fire codes, safety regulations—and performance evaluation, in which a design is assessed for its likely performance in its environment - for example, its aesthetic appeal. The distribution of evaluation criteria between these two categories will have a big impact on the effectiveness of computer-based design tools.

As the randomly-generate-and-test methods lose the purposive feel of design, the question arises as to whether any generative methods can be intent-driven: instead of simply randomly generating a change, assessing its effect, and then deciding whether to accept it, is there a method of proposing synthesis moves oriented towards meeting a goal? Applying a pre-filter to the available moves before generating a change simply brings an implicit, approximate evaluation into play earlier in the process. Indexing or classifying operators according to the expected goals they will contribute to allows move selection to be based on intent, but the problem then becomes one of establishing a useful classification scheme. One type of approach which, it was suggested, does capture this intent-driven aspect of design is that found in case-based and analogical reasoning. Here, a library of known designs is searched and compared to the existing problem. Particular designs are selected, based on their perceived match with the problem. Modifications (moves in a design space) are then made to these designs in order to obtain a full solution to the given problem. Thus the design intent (the problem description) is used to select a set of seed designs. Selecting the appropriate modifications may still require the same approaches as above, but now a localized area of the design space has been selected for investigation.

Note that if the knowledge of the search space is available, then better solutions would probably be found more easily by using intent-based selection of the moves. However, this knowledge is not always available or easy to code, nor is it general. In that case, more general methods, in which all the knowledge is removed from the generation and put into the evaluation may be preferable. Machine learning may be applied to these systems in order to improve their performance on the specific problems, and the resulting move selection techniques would then correspond to intent-based synthesis.

3. Design as Search

The second main area of discussion was the issue of whether design can be adequately characterized as search at all. Again, this hinged in part on the issue of whether human behaviour defines design. A number of points were raised which cast doubt on the adequacy of search as a description of the human design process. It was asserted that architectural designers, for example, tend to pursue a single idea, pushing it to conclusion, and forcing the idea to fit the requirements (or forcing the requirements to fit the idea). If this is the case, the process should not be described as search, as other options are simply not considered. Allied to this is the notion that good design follows from an inspiration, or a "great idea", and that good designs result from adapting that good idea to the present circumstances, rather than vaguely searching through the range of all logical options.

A number of other objections, more specific to formal design, were also raised. There is, in general, no well-defined path from function to form, contrary to the underlying assumption of many prescriptive design methods, and thus design is a matter of choice and judgement, rather than a search for the single answer implied by the problem statement. As discussed in the previous section, designs are frequently not quantifiable, and so search, requiring definite criteria against which different states are compared, cannot be carried out. Even if the designs can be evaluated, the requirements of the problem frequently change, either as a result of finding that the specifications are inconsistent or overly restrictive, or through explicit changes in the requirements or the environment during the process, or through discovering that the problem is under-specified, and that choices have to be made regarding which particular paths will be pursued. Design also involves the phenomenon of emergence: the re-interpretation of a representation producing structures which were not explicitly intended or created, and the use of these emergent structures in later design decisions. Standard search methods give little support for emergence, using fixed representations and interpretations throughout the process. This highlights the need for schemas which can be adapted dynamically as design proceeds.

Contrary to the flow of the above discussion, the question was raised as to whether there can be anything other than search in computer-based design. Almost every proposed approach can be reduced to search, usually by moving up a conceptual level. For example, instead of representing a design task as a problem with changing requirements, we could represent it as a search through a space of (*requirement*, *design*) pairs. However, if everything can be reduced to search, it is then doubtful that reducing processes to search is useful for understanding design. Although characterizing a problem as search might make it easier to implement as a computer system, it hides the difference between different approaches, and obscures their significant characteristics.

Finally, the point was raised that even if all the above objections are accepted, and search is not an adequate characterization of design, that does not mean that search should be abandoned as a tool—many specific design problems can be characterized as search, and many of the alternative approaches would require search as significant sub-processes.

4. Research in Search-Based Design

The third area of discussion during the session was the topic of research methods in search-based design. Firstly, the question was raised as to whether it is advisable to modify existing algorithms (as in the two presentations) to improve results on specific problems, or whether it is better to search for a better representation of the problem, allowing the standard algorithms to be applied successfully. The response was that if the problems are generic, then there is an advantage to modifying the algorithms. Two approaches are possible: designing better algorithms, or designing algorithms which are adaptive, varying their operation according to the specifics of the problem. Related to this topic, it was suggested that as we gather more knowledge on which algorithms perform best on which particular types of spaces, and at what phases of the design process, we should be progressing more towards hybrid algorithms, which start with one method, and then switch to another as we move to a different phase.

A more basic discussion point concerned the purpose of research into design methods. Our efforts should not be restricted to current design practice, but should principally be aimed at producing better designs. Of course, this may well involve providing better support for current practice, but it may also involve radically different methods, either in producing autonomous design systems, or using computer-based tools to change the way in which designers design. For example, it was mentioned above that architectural designers typically only pursue a single idea. Why is this? Is it because this is the best way of producing good designs, regardless of the support available, or is it a result of the limitations of human information management? It could be that managing multiple alternatives to any significant level of detail, including diagrams, models, evaluations, and ramifications is too complex a task for unsupported humans, but if support for these aspects was provided, and all the information made readily available, the standard design process might change from pursuing a single option to pursuing multiple options. Such questions need to be investigated, in order to direct future research towards the goal, as stressed above, of producing better designs.

As a related point, we should consider whether or not "great idea" design is a good exemplar for our research. On a number of occasions during the discussion approaches were evaluated by comparing them to the approaches used to design well-known objects—for example, the Sydney Opera House. It is not clear that such examples provide a good focus for design computing research. In many ways, the successes or failures of these projects are discussed in terms of their uniqueness, the way in which they shift the bounds of current taste, and their innovative approach, and not necessarily in terms of how well they meet their original purpose. These criteria depend on an understanding of human reactions to buildings, human preferences, and attitudes to art and aesthetics. If our aim is to produce better designs, then our efforts should perhaps be directed towards more commonly encountered design tasks. Research would thus be centered on increased automation of design tasks which can be readily evaluated, and on generating alternatives and providing bookkeeping support for these more speculative tasks.

Finally, it is worth noting the seeming change in focus of design research between the workshop discussed here and the previous one in the series (Gero and Tyugu, 1994). In that previous workshop, there were six papers explicitly concerned with generative systems (Carlson, 1994; Heisserman and Woodbury, 1994; Shih and Schmitt, 1994; Brown et al., 1994; Andersson, 1994; Cagan and Mitchell, 1994). All six of them considered grammatical systems in particular. All except Cagan and Mitchell were largely concerned with presenting the spaces and the formalisms, with little concern for how those spaces would be searched. In the current workshop, there were five papers explicitly concerned with generation and search (the first three from this session, plus Maher et al. (1996) and Gero and Kazakov (1996)). All five of these papers focussed on methods of searching the spaces, and further, all five used non-systematic methods to carry out that search. Although conclusions should not be drawn from such a small sample, it would appear to indicate a development in the field, in that the research is now building on a framework laid down earlier, and considering how that framework can be used in practice.

5. Conclusion

Very little consensus was reached during the discussion. One point worthy of note was the acceptance by almost all present of the idea of exploration being a better description of the design process than pure search. Other than that, the discussion served to highlight that there are many questions which remain unanswered, both in the general area of relating the design process to generation, and in specific issues in the control of the search and generation process.

Acknowledgments

I would like to thank Stephan Rudolph, for providing the clear and comprehensive notes on which this summary is based, and Mary Lou Maher, who chaired the discussion and gave it its structure by posing the three questions to be discussed. Jon Cagan provided a number of helpful comments on an earlier draft. Finally, I must thank all the participants of the workshop, who were responsible for most of the ideas raised in this summary; I apologize for not providing individual attributions.

References

Andersson, K.: 1994, A vocabulary for conceptual design, *in* J. S. Gero and E. Tyugu (eds), *Formal Design Methods for CAD*, North-Holland, Amsterdam, pp. 157-171.

Brown, K. N., McMahon, C. A. and Sims Williams, J. H.: 1994, A formal language for the design of manufacturable objects, *in* J. S. Gero and E. Tyugu (eds), *Formal Design Methods for CAD*, North-Holland, Amsterdam, pp 135-155.

Brown, K. N. and Cagan, J.: 1996, Modified shape annealing for optimally-directed generation: Initial results, *in* J. S. Gero (ed.), *Advances in Formal Design Methods for CAD*, Chapman & Hall, London, pp. 59-73.

Cagan, J., and Mitchell, W. J.: (1994, A grammatical approach to network flow synthesis, *in* J. S. Gero and Tyugu, E. (eds), *Formal Design Methods for CAD*, North-Holland, Amsterdam, pp. 173-189.

Carlson, C.: 1994, Design space description formalisms, *in* J. S. Gero and Tyugu, E. (eds), *Formal Design Methods for CAD*, North-Holland, Amsterdam, pp. 121-131.

Gero, J. S. and Kazakov, V. A.: 1996, Evolving building blocks for design using genetic engineering: A formal approach *in* J. S. Gero (ed.), *Advances in Formal Design Methods for CAD*, Chapman & Hall, London, pp. 31-50.

Gero, J. S. and Tyugu, E. (eds): 1994, *Formal Design Methods for CAD*, North-Holland, Amsterdam.

Heisserman, J. and Woodbury, R.: 1994, Geometric design with boundary solid grammars, *in* J. S. Gero and Tyugu, E. (eds), *Formal Design Methods for CAD*, North-Holland, Amsterdam, pp. 85-105.

Kirkpatrick, S., Gelatt, C. D. Jr and Vecchi, M. P.: 1983, Optimization by simulated annealing, *Science*, **220**(4598), 671-679.

Maher, M. L., Poon, J. and Boulanger, S.: 1996, Formalizing design exploration as co-evolution: a combined gene approach, *in* J. S. Gero (ed.), *Advances in Formal Design Methods for CAD*, Chapman & Hall, London, pp. 3-30.

Simon, H.: 1967, *The Sciences of the Artificial*, MIT Press, Cambridge, MA.

Shih, S.-G. and Schmitt, G.: 1994, The use of post interpretation for grammar-based generative systems, *in* J. S. Gero and Tyugu, E. (eds), *Formal Design Methods for CAD*, North-Holland, Amsterdam, pp. 107-120.

Smithers, T., Corne, D. and Ross, P.: 1994, On computing exploration and solving design problems, *in* J. S. Gero and Tyugu, E. (eds), *Formal Design Methods for CAD*, North-Holland, Amsterdam, pp 293-313.

Zozaya-Gorstiza, C. and Estrada, L. F.: 1996, Incorporating heuristics and a meta-architecture in a genetic algorithm for harness design, *in* J. S. Gero (ed.), *Advances in Formal Design Methods for CAD*, Chapman & Hall, London, pp. 75-96.

PART THREE

Performance Evaluation Methods in Design

7

A PERFORMANCE-BASED PARADIGM OF DESIGN

YEHUDA E. KALAY
University of California, Berkeley, USA

AND

GIANFRANCO CARRARA
Università degli Studi di Roma La Sapienza, Italy

Abstract. This paper proposes an alternative approach to existing design paradigms (and their CAD implementations) that are based on the traditional, causality-based notion that 'Form follows Function.' The proposed approach, which will be referred to as *performance-based design*, is founded on the argument that the relationship between form and function is *contextual* rather than *causal*. Hence, the expected *performance* of a given design proposal can only be determined by an *interpretive evaluation*, which considers the form (and other physical attributes) of the proposed design, the functional objectives (goals) that it attempts to achieve, and the circumstances under which the two come together (the context). The paper develops a performance-based design methodology and demonstrates its application in an experimental, knowledge-based CAD system.

1. Introduction

The quest for understanding how humans perform complex cognitive activities such as architectural design, and for developing methods and tools that can help them consistently achieve desired results, has been the raison d'être of design methods research for the past four decades. The formulation of such methods has followed, by and large, the scientific method of developing *theories* that explain the process of design, then casting them in *models* that can be represented explicitly and *implemented* by computational and other means. This endeavor has been mostly guided by the conventional wisdom that architecture, more than any other design activity, is a process that seeks a convergence of *form* and *function* within a particular spatio-temporal *context*: a physical container, or a stage, which will support certain human activities, subject to certain conditions and constraints, while being embedded in a particular social and physical context.

Following Thomas Kuhn's influential work concerning the nature of scientific inquiry (Kuhn, 1962), many design methods researchers sought a *causal* relationship between form and function, and a method for deriving *form* from *function*, or *function* from *form*. At the core of this quest lay two assumptions:

1. That a physical system's significant geometrical (and material) properties have some functional utility, and that one form is more suitable to fulfilling that function than other, alternative, forms.
2. That finding a causal relationship between form and function will lead to the development of a method, which can be applied with some assurance of success in every case where a form must be produced that will optimally facilitate and support a given set of functional needs, or that the function of a given object can be determined by closely examining its form.

Two fundamentally different paradigms of design have emerged, representing two different interpretations of the causal relationship between form and function. The first, attributed to Simon, Newell, and Shaw (Simon, 1979), attempts to explain design as a unique instance of general *problem solving*. It postulates that designers start with the sought *function* (i.e., the desired behavior of the system), which is often represented as a set of goals and constraints. The designers then attempt to discover a *form* that will support the desired function, using *deductive* search strategies.

The other paradigm, which is called *puzzle-making*, has emerged from the work of researchers like Alexander (1964) and Archea (1987). It postulates that designers begin with a set of *forms* (that include materials as well as geometry), which are modified and adapted until they achieve some desired *functional* qualities. This paradigm is based on *inductive* reasoning, and has been modeled with the aid of analogical inferencing methods (metaphors, symbols, and case studies).

While logically consistent and computationally convenient, neither one of these two paradigms, nor their many derivatives and permutations, has gained favor with architects themselves. The essence of the profession's criticism can be summarized as the failure of the proposed paradigms to account for the celebrated *intuitive leap*, that elusive but well-known moment when form and function seem to converge into a meaningful whole (Norman, 1987). From a logical point of view, the intuitive leap represents a *discontinuity* in the causal relationship between form and function, hence a stumbling block in developing a uni-directional (form-to-function, or function-to-form) design theory.

This paper proposes an alternative approach to understanding the process of architectural design, which attempts to account for the 'intuitive leap' experience and fit it within a design paradigm that can be formally modeled and explicitly represented. The proposed approach, which will be referred to as *Performance-Based Design*, does not attempt to formalize the intuitive leap itself, only to *accommodate* it in the paradigm. It proposes to do so by promoting *interpretive evaluation* over *causality*. Instead of deriving form from function, or function from form, it advocates the development of means that measure *performance*–the interpreted utility derived from a given form and a given set of functional requirements within a particular spatio-temporal context.

The notion of *performance* is derived from the argument that the relationship between form and function is *context-based*, rather than *causality-based*. That is, the performance of a proposed design solution can only be determined by an interpretive evaluation which considers the *form* (and other physical attributes) of the proposed solution, the *functional* objectives (goals) that it attempts to achieve, and the *circumstances* under which the two come together. *Performance-based design* recognizes that different forms can successfully achieve similar functions, and that different functions can often be derived from similar forms. In addition, it accounts for performance variances of the same form/function combinations within different contexts. Consider the following example:

> If all the chairs in a lecture hall are occupied, students will often sit on the floor. The shape (form) of the floor is not at all similar to the form of the chairs (nor is it usually made of the same materials). Yet, under certain circumstances and within certain limits, it is *functionally equivalent* to chairs. The same floor can also be used for many purposes other than sitting. For example, it can be walked upon, danced upon, and large objects can be placed on it. It can even be used to identify groups of spaces (as in 'first floor'). Included among all these varied functions is also the function of sitting. Obviously, the floor is not as well adapted for the function of sitting as are chairs, and most students will prefer to sit on chairs whenever possible. However, if chairs are not available, they will, in many cases, consider the floor *functionally adequate* for the purpose of sitting. The floor will *not* be considered functionally adequate for sitting when the context is an elegant cocktail party in the White House, as compared to a lecture in a university, and when the participants are foreign diplomats and not foreign students. In that case, the participants will prefer to stand, even if standing causes them considerable discomfort.

A computational implementation of this paradigm, which considers form and function to be equally important, requires that both be represented explicitly, along with the context of the particular design problem. The 'adequacy' of a particular combination of form-function-context will then be determined by evaluators (which may very well include the designer himself), which use a variety of means to predict the performance that will ensue from the combination of form, function and context, then judge the desirability of that performance in comparison with alternative form-function-context combinations.

In the following pages, we develop the argument for performance-based design. Then we present a model of computer-aided performance-based design, and present an experimental knowledge-based CAD system that implements it. Initially, we will use the terms 'form,' 'function,' 'context' and 'performance' loosely, relying on the reader's intuitive understanding of their meaning. We will define these terms more precisely in the second part of the paper, where such rigor is needed.

2. Causality-Based Design Paradigms

The notion that Form follows Function is derived from the assumption that a system's significant geometrical (and material)[1] properties have some functional utility, and that one form is more suitable for fulfilling that function than other, alternative forms. This notion, and its inverse (function is derived from form), have guided architects and engineers for millennia. Among other achievements it has also provided a convenient causal relationship between *form* and *function*, the two pillars of architectural design, hence for developing theories and methods intended to assist architects in performing their increasingly more demanding task of finding the "right" form-function combination.

Many formal theories and methods that were developed over the years to assist architects have been based on this logical foundation. They can, nonetheless, be classified into two general groups (Aksoylu, 1982): those that follow Simon, Newell and Shaw's (1979) Problem-Solving paradigm, and those that follow what Archea (1987) called "Puzzle-Making."

2.1. PROBLEM-SOLVING

Problem-solving is a general theory that attempts to explain the cognitive process of creative thinking. It was first formalized by Simon, Newell and Shaw in the late 1950s, and implemented in a computer program called GPS (General Problem Solver). Problem-solving assumes that the desired effects of some intellectual effort can be stated in the form of constraints and goals at the outset of the quest for a solution to achieve them. To find the solution, the problem solver uses a variety of search strategies to generate successive candidate solutions and test them against the stated goals, until one is found which meets them. The goals thus 'guide' the search for a solution right

---

[1] In the following, the term 'form' will be used to refer to all the physical attributes of objects, including their material composition, surface finish, and so on.

from the beginning of the problem-solving process. Problem-solving assumes that setting goals (i.e., knowing what should be accomplished) can be separated from the process of finding a solution that meets them, and that such knowledge can be acquired through an independent inquiry (analysis), which should be completed before the search for a solution has been initiated (Akin, 1978). For example, using this approach, selecting a structural system to span some opening will generally follow after an analysis of forces, cost, and other characteristics of the structure have been determined.

Since the characteristics of the problem, according to the problem-solving paradigm, are known prior to commencing the search for the solution itself, its proponents hold that the search for a "satisficing"[2] solution is goal-directed, and therefore that means-ends analysis can be employed to guide the search towards finding the desired solution. Thus, the skills that are employed when following the problem-solving paradigm are mainly *analytical*: the ability to compare the current "state" of the designed artifact to its desired "state" (in terms of its expected utility and behavior), and the ability to draw operational conclusions from this comparison so that the differences can be reduced.

Such goal-driven approaches have been computationally represented as deductive, backward-reasoning search strategies, where operators are applied to the goal statement in order to convert it into a set of sub-goals that are easier to solve. This method is applied recursively until a set of subgoals that can be solved without further reduction is found (Laird et al, 1986). Examples of tools based on this paradigm include space allocation programs (Armour and Buffa, 1969; Shaviv and Gali, 1974), and a large number of evaluation programs, such as wayfinding and energy (Gross and Zimring, 1992; Shaviv and Kalay, 1992).

2.2. PUZZLE-MAKING

The assumption that, in architecture, the characteristics of the desired solution can be formulated prior to and independently of the search for the solution that satisfies them was rejected by critics like Archea (1987) and Bijl (1987). They argued that such knowledge cannot exist prior to the search itself, since the sought solution is *unique*, and the process of finding it is characterized by *discovery* and has to contend with *uncertainty*. Kim (1987) and others have argued that the brief that architects are given by their clients, and which often constitutes the basis for the design goals, is much too vague, in most cases, to form a complete goal statement. Rather, it is merely a

[2] Meaning "good enough." The term was coined by Herbert Simon in his *book Sciences of the Artificial*, MIT Press, 1969, pp. 35-36.

statement of *intents*, which defines a general framework for the sought solution, and some of the constraints it must abide by. Instead, they suggest that architects must gradually develop the statement of goals as they proceed with the design process itself. The additional information needed to complete the goal statement must either be *invented* as part of the search process, or *adapted* from generalized precedents, prototypes, and other relevant past experiences (so-called 'design cases'). Since the relationship between the newly invented information, as well as the precedents, to the particular needs of the problem can be discovered only as the problem becomes clearer, the adaptation itself is problem-specific, and cannot be accomplished prior to engaging in the search process itself.

Design, according to this view, is a process of *discovery*, which generates new insights into the problem. The design search process may, therefore, be compared to *puzzle-making*–the search for the most appropriate effects that can be attained in unique spatio-temporal situations through the manipulation of a given set of *components*, following a given set of combinatorial *rules*. Since architects cannot invent information from scratch in every case, they rely on design 'cases,' either from the architect's own experience or from the experience of the profession at large, to provide them with a rich pool of empirically validated information which has been refined through many years of practice and has gained society's or the profession's approval. This information comes in the form of proven solutions (Alexander et al, 1977), architectural styles, celebrated buildings, established metaphorical relationships, and recognized symbolisms (Venturi, 1972). How architects adapt this body of knowledge to the particular problem at hand is not known–it is the essence of architecture's celebrated 'intuitive leap' and creativity.

Therefore, rather than rely on a goal-driven strategy, the puzzle-making paradigm relies on adaptation of precedents, symbols, and metaphors. The main skills employed when following this paradigm are *synthetical*: the ability to compose given parts into a new, unique, whole. Such *data-driven* approaches have been computationally represented as forward-reasoning search strategies: operators are applied to the current state of the problem with the aim of transforming it according to pre-set rules. Example of tools based on this approach include generative expert systems, shape grammars, and case-base design systems (Coyne et al, 1990; Flemming, 1994; Oxman, 1992).

3. Different Kinds of Relationships between Form and Function

In this paper we argue that the relationship between Form and Function is much more complicated than implied by the causality-based notion of

'Form Follows Function,' and its inverse. Indeed, a particular *form* is often capable of many different functions, and a similar *function* is often afforded by many different forms. The following examples will serve to illustrate this argument.

3.1. MANY FORMS, SAME FUNCTION

The over-simplicity of the notion 'Form Follows Function' is evidenced by the multitude of different forms which essentially were designed to support similar functions.

Chairs provide one of the best examples of different forms that were developed to support exactly the same function (sitting). Figure 1 shows some of the many designs of dining/table chairs.

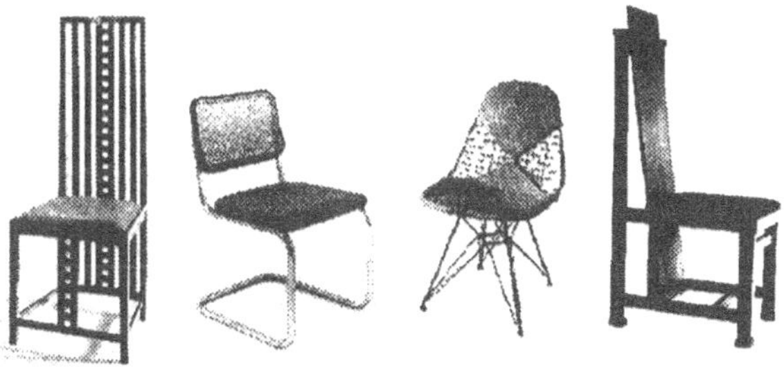

Figure 1. Chairs demonstrate how different forms can support the same function.

Design competitions, where competitors must respond to the same set of functional requirements within the same context, provide additional evidence that in architecture, form does not necessarily follow function. Each and every competitor will, invariably, produce a very different form.

Some scholars have tried to explain this apparent lack of causality by arguing that, typically, the functional requirements of a building do not tightly constrain its form, thus leaving the architect with much room to entertain 'styles' and other 'non-practical' considerations. Herbert Simon, for example, has defined style as "one way of doing things," chosen from a number of alternative ways (Simon, 1975). Since design problems generally do not have unique or optimal solutions, says Simon, style can be used to select a solution from among several *functionally equivalent* alternatives, just as any other criteria can. He offers the following analogy:

> Mushrooms can be found in many places in the forest, and the time it takes us to fill a sack with them may not depend much on the direction we wonder. We may feel free, then, to exercise some choice of path, and even to introduce additional choice criteria ... over and above the pragmatic one of bringing back a full sack (of mushrooms).

Most architects, however, would reject this notion that form is the result of less 'practical' functional considerations than other aspects of the building, and therefore an afterthought, something to be contemplated only when all the other 'important' aspects of the design have been dealt with. Rather, they would argue, it is something a competent architect will consider *before*, *during*, and *after* the development of solutions satisfying the functional needs. Moreover, the two issues cannot be separated, since each one informs the other, and influences its development.

3.2. MANY FUNCTIONS, SAME FORM

The notion that a given form can support many different functions is demonstrated well by designs of playgrounds, parks, and civic plazas. Joost van Andel (1988) observed that playgrounds for children between the ages 3 and 7 function best if the activities they afford are less structured, in terms of the equipment they contain. For instance, placing an old fire engine in a playground (a form) will direct the children's activities towards particular play patterns. Furthermore, van Andel observed that this particular form tends to create gender-biased play patterns, which appeal more to boys than to girls. On the other hand, a playground which consists mostly of a sandbox, some rocks, and a few trees or bushes affords less restricted play patterns, and is equally accessible to both boys and girls. He attributes this performance to the creative imagination of the children, who can translate the existing, generic forms into particular ones, as needed for playing games such as 'house,' 'cops and robbers,' or the landing of an alien spaceship.

Another example of architectural multipurpose (i.e., functional) spaces has been described by Elizabeth Cromley in her paper on the history and evolution of modern bedrooms (Cromley, 1990). In addition to providing a place for sleeping, bedrooms through the seventeenth century also functioned as parlors, dining rooms, and as places for entertaining guests. In the eighteenth century, the function of bedrooms became more focused, as a place for sleeping and dressing, for quiet retirement, and for socializing with close friends and family members. In the nineteenth century, bedrooms became a place to occupy only at night. In the 20th century, the definition of their function was broadened again, especially as far as children's bedrooms were concerned. Today such functions include sleeping, doing homework, reading, and playing with friends. Bedrooms for the adults (the so-called 'Master Bedroom'), have turned into 'suites,' which include full bathrooms, dressing rooms, and walk-in closets.

The ability of the same form to afford different functions is further demonstrated by what we now call *adaptive re-use*. The term designates the conversion of older buildings to meet modern needs. It is rooted in the economic realities of the late 20th century, and the growing need for urban renewal and rehabilitation. This trend is characterized by corporations, shops, and even residential units moving into older buildings in the core of cities. Rather than tear down a building which may have some historical or cultural significance, new tenants may rehabilitate it while preserving its character. A typical case in point is Adler and Sullivan's Guarantee building in Buffalo, New York (Figure 2). In its 100 years history, this landmark building has served successfully as an office building, department store, and a variety of commercial, government, and other functions. While the building's interior has undergone some changes, the building has remained largely intact.

Figure 2. Same form, many functions: the Guarantee building, in Buffalo, New York.

3.3. OTHER KINDS OF FORM-FUNCTION RELATIONSHIPS

Peter Eisneman's structuralist approach to buildings, which derives from his own interpretation of Noam Chomsky's linguistic theories (as well as Jacques Derrida and other philosophers), demonstrates well the complexity of the possible relationships between form and function. His proposed design for the Max Reinhardt Haus in Berlin (a cultural center and office building), is modeled as a huge, three-dimensional Möbius strip–an abstract topological construct (Figure 3).

Figure 3. Arbitrary form: Peter Eisenman's proposed design for Reinhardt Haus, Berlin.

3.4. THE IMPORTANCE OF CONTEXT

The form of a building also depends upon the physical, cultural, social, and other contexts in which it is embedded, at least as much as it depends on the function it must serve. The form of the Sydney Concert Hall, which is depicted in Figure 4, is an example of a form derived from the physical context of the building (the Sydney harbor), as much as from its function (a symphony hall).

Figure 4. Context-influenced form: the Sydney Opera House.

Likewise, the shape of Le Corbusier's Ronchamp Chapel has been derived from its spiritual context, as much as from its functional and physical site considerations; and Gerrit Rietveld's colorful Schröder House in Utrecht, The Netherlands (1931), has been shaped as much by the neoclassicist cultural ideas of the De Stijl movement to which he belonged, together with painters like Theo van Doesburg and Piet Mondrian, as much as by functional requirements.

3.5. THE RELATIONSHIP BETWEEN FORM, FUNCTION AND CONTEXT

The position taken in this paper is that Form, Function and Context are linked through tri-lateral, mutual interdependencies. It would be futile, however, to look for causality among these relationships. The utility of the links can only be revealed by observing, measuring, and interpreting their overall, combined result, which is what we call *performance*. Performance evaluation is intended, therefore, to assess the desirability of the confluence of the three factors. It may reveal, for example, that a particular form is capable of supporting a certain functional need in a particular context, in which case it will be deemed 'successful.' On the other hand, it may reveal a need to modify the form to meet the desired function in the particular context, or to modify the desired function to meet the functionalities afforded by that form in that particular context.[3]

The complexity of this relationship is exacerbated by the fact that the *nature* of Form, Function and Context are dissimilar. Functional objectives are often *abstract*, expressed in terms of social, psychological, economic, and other behaviors. Forms, on the other hand, are often quite *specific*, and are expressed in terms of topology, geometry, and materiality. Context typically involves given social, cultural, and economic situations, in addition to physical ones (which include topography, climate, flora, etc.). To bridge the representational gap between the functionalities afforded by a given form, or the form that will afford a particular function in a particular context, the designer must rely on an *interpretive-evaluative* process, with its intrinsic fuzziness, value-laden biases and subjective belief systems. He must attempt to *predict* the functionality (behavior) afforded by the chosen form within the prescribed context, thus translating the form into a functional abstraction. At the same time, he must *envision* the form that might afford the sought function, thus translating the functional abstraction into a physical form. He must then *compare* the two abstractions to determine if they match (Figure 5). In addition, he must map the emerging composition onto the context, to determine if they match it too.

---

[3] In some cases it may also be possible to modify the context itself, for instance by obtaining exemptions to zoning codes, modifying the socio-economic makeup of the inhabitants, or even the physical characteristics of the site.

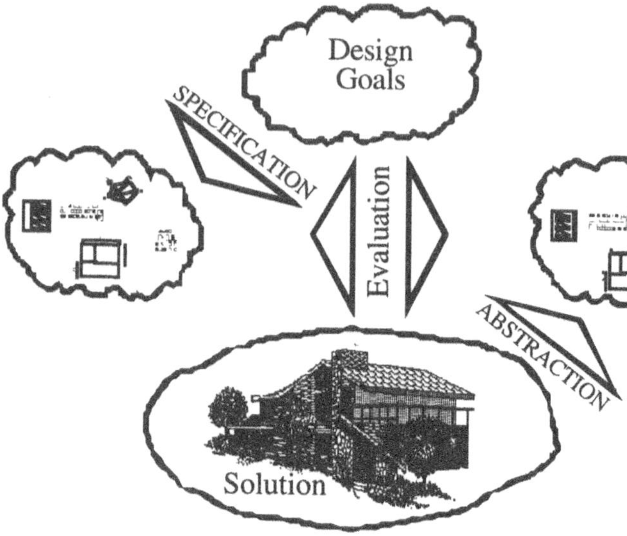

Figure 5. Specification and abstraction in the design process.

Another complication arises from the multiplicity of *functions* that must be supported by the same form, often at the same time, as well as the multiplicity of *forms* that are needed to support a given function. Windows in a building offer a good example: they are needed to admit light and view into the building, but they interfere with the shelter-providing functions of the walls they puncture (from thermal, humidity, air infiltration, and other environmental control points of view). At the same time, they also contribute to the aesthetic appearance of the building, where their shape, pattern and rhythm may or may not correspond to their function. Consider, for example, the two functionally-equivalent, but aesthetically-different views of Le Corbusier's Villa Savoye (Figure 6). The task of the designer, in many cases, is to find a good compromise of form, function and context, rather than an optimal relationship between them (which may be the goal in certain engineering fields).

4. Performance-Based Design

Having denounced the causal relationship between Form and Function, which, as argued earlier, is the basis for the prevailing design paradigms, we must find another paradigm that will explain how designers can bridge the gap between Form and Function, and are thereby able to justify the selection of a particular Form to meet specific a Function within a particular Context.

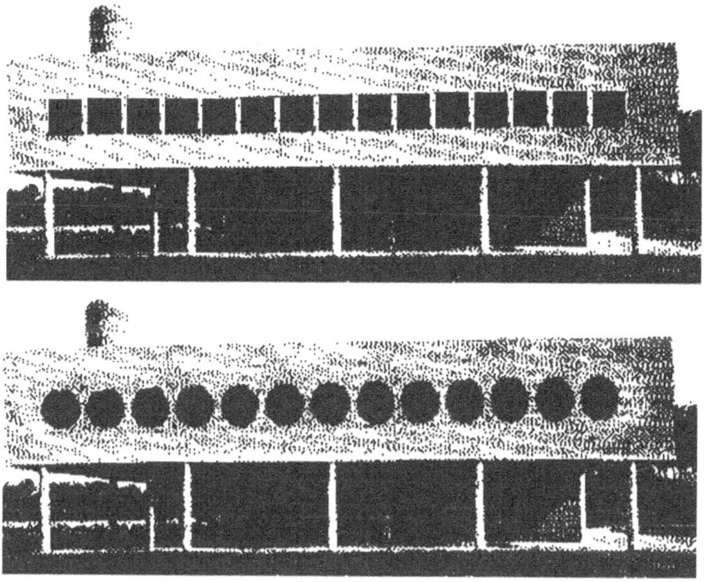

Figure 6. Functionally-equivalent, but aesthetically-different views of Villa Savoye.

This paper suggests that such a paradigm can be formulated. We call it *performance-based design*: the specific confluence of *form* and *function* in a particular *context* (Figure 7). We define Function as the desired *behavior* of the building (or other artifact). This behavior can be quite specific (e.g., budget), or more abstract (e.g., provide a conducive environment for work). We use Form in its conventional connotation, as a physical manifestation of topology, geometry, and materiality. By Context we mean the physical, social, economic, cultural, legal, and other *settings* and *events* in which the building is located. By comparing the physical manifestation of a given building form with the conditions necessary to fulfill a desired set of behavioral characteristics of that building, within the particular context in which it is situated, we can determine the *performance* of both its form and function, relative to another composition of form and function within the same context.

This design paradigm also explains (and rationalizes) why designers may begin the search for a form-function-context composition either with a given function or with a given form, while progressing towards the other, as well as skipping around, back and forth, between the two (Figure 8). D*esigning*, accordingly, can be considered an iterative process of search, where desired functional traits are defined, forms are proposed, and a process of evaluation is used to determine the desirability of the performance of the confluence of forms and functions within the given context (Carrara et al, 1994). The search terminates when the designer finds a form that fulfills the function, or is satisfied by the functionalities afforded by the chosen form, within the given context. We call this condition *functional adequacy*: the instance when form and function come together to achieve acceptable *performance* within a given context.

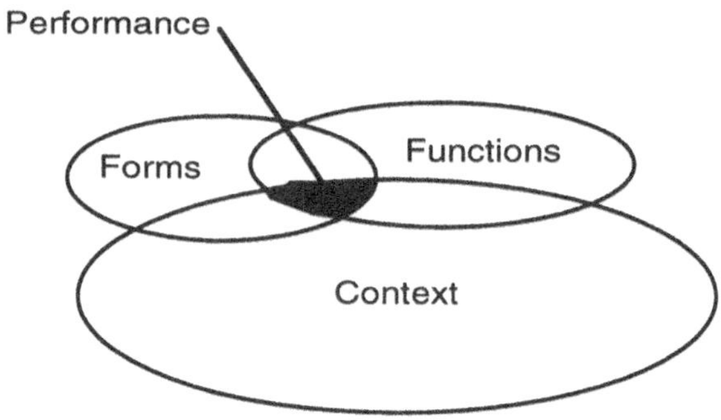

Figure 7. Performance, as the confluence of form, function and context.

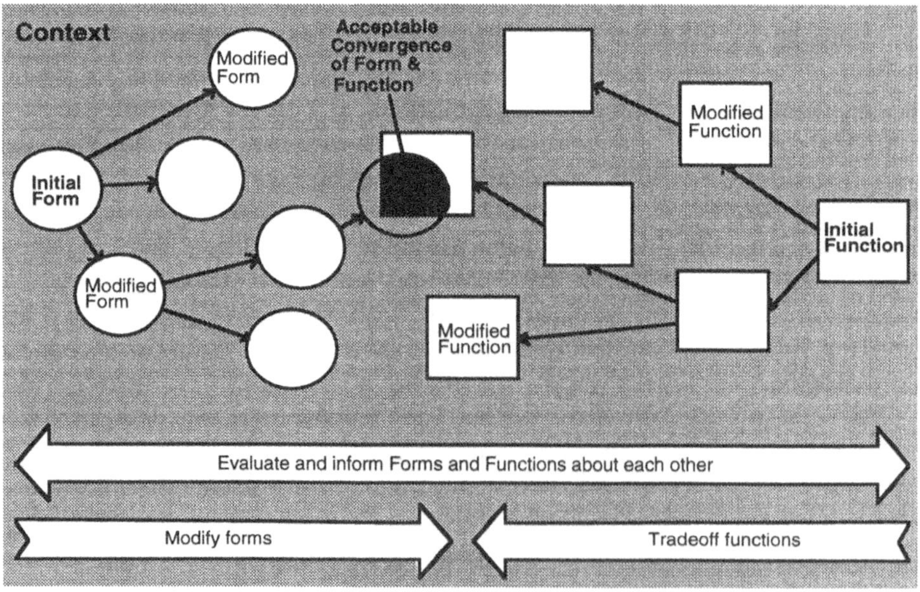

Figure 8. Design as a bi-directional search.

5. Implementation

To implement the proposed paradigm, a computable model of performance-based design had to be developed, which can represent explicitly *function* and *form,* as well as the *context* of the particular design project. Additionally, for the sake of computability, the *process* that helps to identify an acceptable convergence of form, function and context also had to be represented explicitly.

Through a succession of projects, beginning in 1985, we have developed such a model, which consists of four strongly related components:

1. A structured set of **goals**, representing the functional requirements that a design solution must meet.
2. A structured set of **solutions**, representing the physical and spatial components of the emerging solution (rooms, walls, windows, etc.).
3. A representation of the **context**, in which the project is embedded (physical, social, cultural, etc.).
4. A structured set of **evaluators**, whose purpose is to predict the performance of the form-function-context composition based on the physical attributes of the objects and the goals, within the particular context of the design problem.

These four components rely on different methods of representation. Goals are represented by sets of functionally equivalent *constraints*. Solutions are represented by specific *building elements* (e.g., walls, spaces, and materials). Contexts are represented through constructs we call *settings* and *events*. Evaluators are represented by a variety of methods, including simulation, case-based knowledge, and other computational means.

5.1. REPRESENTATION OF DESIGN GOALS

Many (but obviously not all) functional needs can be represented as objective *constraints*. For example, the nature, morphology, and sizes of spaces, their material composition, the equipment and furniture used in them, and the procedures for managing them can be represented as a class of *use* constraints. Desired temperatures, humidity, lighting, and other comfort parameters can be represented as a class of *environmental* constraints. The *structural* and *mechanical* behavior of buildings comprise additional sets of constraints, as do their behaviors under exceptional conditions such as fire and earthquakes, which can be represented as a class of *safety* constraints. Such classes can be further divided recursively into sub-classes, creating a hierarchical structure of increasingly more specific and detailed constraints (Figure 9). The classifications are, of course, highly arbitrary, and can be tailored to the needs and preferences of each designer. They demonstrate, nonetheless, the ability to define design objectives explicitly, in terms of testable sets and subsets of goals and constraints. Some of these objectives are derived from the clients preferences (e.g., budget). Others may be derived from the context (e.g., views, temperature, size limitations). And still other objectives may reflect the architect's own aspirations, style, and ethical code.

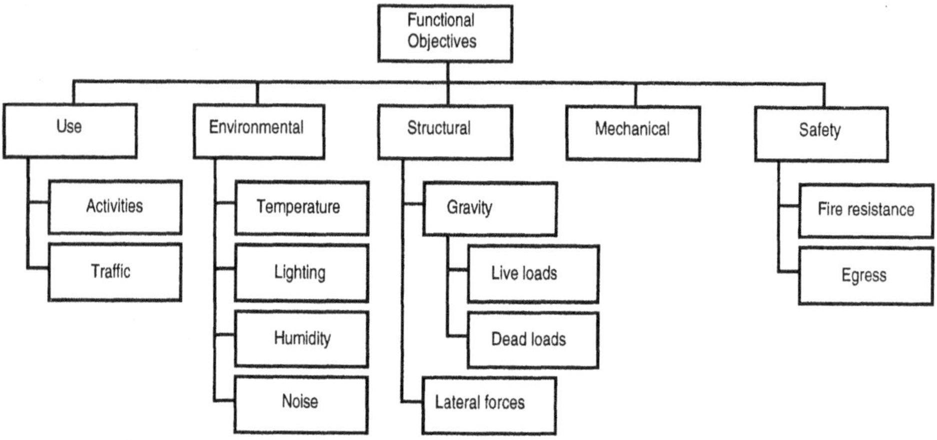

Figure 9. Hierarchical, recursive relationship between goals and constraints.

A set of constraints can be used to indicate a particular combination of desired behaviors that must be accomplished by a candidate design solution in order to achieve a specific design objective. We call this set a *goal*. For example, the number of bedrooms in a single-family house determines the number of bathrooms it should have, because it is often indicative of the size of the family that will occupy the house. Likewise, the adjacencies of the rooms cannot be separated from their number and the types of activities they contain. The number of rooms and total floor area is directly related to the budget, which is also influenced by the quality of the construction.

The goals are considered to be achieved if all their constraints have been satisfied. The particular combination of constraints that is considered a goal is established when the goals are first introduced. This forces the designer (and the client) to consider and establish reasonable combinations of objectives, which then guide the design process. Additional goals may be added, or existing goals may be modified or deleted during the design process, thereby providing a measure of flexibility and a means for representing changing preferences as the design evolves.

The specificity of design goals must not be confused with the specificity of the design solutions that satisfy them. As argued earlier, different design solutions may achieve the same goal, albeit each may satisfy the constraints that comprise the goal differently. The different performance levels at which alternative sets of constraints may be satisfied represent *tradeoffs* in the context of achieving a particular goal. The different windows on Le Corbusier's Villa Savoye (Figure 6) demonstrates such trade-offs with respect to aesthetical and functional considerations, although both provide the same basic utility of light, ventilation, and view.

While alternative goals represent acceptable combinations of performance levels, some combinations may be preferable to others. A prioritization of goals, reflecting a descending order of preferences, may be imposed by the designer or by the client, indicating which combination of performances the designer should attempt to accomplish first. Such prioritization is not only a common practice when architects and clients are faced with limited resources, but it also has a very profound effect on the direction of the design search process and on its results. This is due to the fact that all the decisions leading to the specification of a design solution are connected to each other, and decisions made earlier in the process may limit the options available to the designer in later design phases, sometimes to a degree where no options are available at all. For example, choosing a particular construction method early in the design process (e.g., wooden frame) imposes many constraints on the building, limiting the options available to the architect in designing its form, details, and construction schedule.

5.2. REPRESENTATION OF DESIGN SOLUTIONS

The stated constraints can be achieved by different, yet functionally equivalent solutions, comprising building *objects*. Computationally, objects can be defined in many ways.[4] Recently, frame-based, object-oriented methods have been gaining popularity. In addition to their computational advantages, object-oriented programming methods appear to be intuitively similar to the building objects they represent. Frames make it possible to encapsulate many of the attributes constituting an object, and they can be organized into hierarchical classes and other types of relationships according to their properties.

According to the frame formalism, the relationship between an object and its attributes is fixed. The *values* of the attributes themselves, however, are not fixed: they are *variables*. Such fixed-attribute, variable-value

---

[4]See, for example "Computer Integration of Design and Construction Knowledge" (Eastman) and "Intelligent Systems for Architectural Design" (Watanabe), in *Knowledge-Based Computer-Aided Architectural Design* (1994), Carrara & Kalay, eds., Elsevier Science Publishers, B.V., Amsterdam.

relationships are known as *name-value* pairs. Attributes (also known as *slots*) can be thought of as 'place holders,' or as predefined properties that are associated with particular types of values. Values (also known as *fillers*), include the permissible range of numbers (and other types of values) that can be associated with a particular attribute, defaults, and even instructions (so called 'demons') that allow the attribute to calculate its value when it depends on values associated with other attributes. For physical objects of the kind used in architectural design, there would typically be an attribute called *shape*, whose value would be a particular topological/geometrical entity describing the form of the object and its location in space relative to some frame of reference. There would also be attributes for material composition, structural properties, cost, thermal properties, and so on.

Every type of entity in the database is accompanied by operators that can create, delete, and modify it, and associate it in various types of relationships with other entities. These operators help to maintain the consistency of the information in the database, by propagating changes caused by outside operators (e.g., by the designer). We have chosen to link objects with four types of relationships, as depicted in Figure 10. They include:

- *Classification relationships*, which provides the means for associating individual objects with classes of objects of the same kind, such that shared properties can be inherited along generalization hierarchies.
- *Instantiation relationships*, which provide the means for making *instances* of a *template* object, and maintain a measure of control over the instances by automatically changing them when certain key attributes of the template change.
- *Assembly relationships*, which connect instances to each other through links that form part-whole hierarchies. Assembly hierarchies allow propagation of changes from 'parent' objects to their dependent 'children.'
- *Aggregation relationships*, which bind together objects that share some common property, or objects that must be considered together when some database changes occur, but do not fall into one of the other categories. Aggregation relationships require explicit definition of the nature of the link that connects them. This explicit definition makes aggregation a more general type of relationship than classification and assembly relationships, where the nature of the link is implied by the type of the link itself. By adding a conditional component to the definition (in the form of a set of rules), the aggregation relation can exercise the link selectively, depending on the nature of the change and the nature of the affected objects.

5.3. REPRESENTATION OF CONTEXT

We consider *context* to comprise of project-independent information that the architect must respond to in his design, and over which he has little or no control. For example, the topography of the site, its climate and views, are such information. Likewise, the cultural environment, the economic and political makeup of the society in which the project is embedded, and often building and zoning codes, comprise project-independent factors that the building must respond to. Additionally, we also include in the term *context* the predominant activities that the building (or urban place) must support, which are typically implied by the nature of the project itself. For example, the medical procedures for treating patients in a hospital, the method of teaching in a school, and even the traditional habits of a family within its own house, are factors the architect must account for in his design.

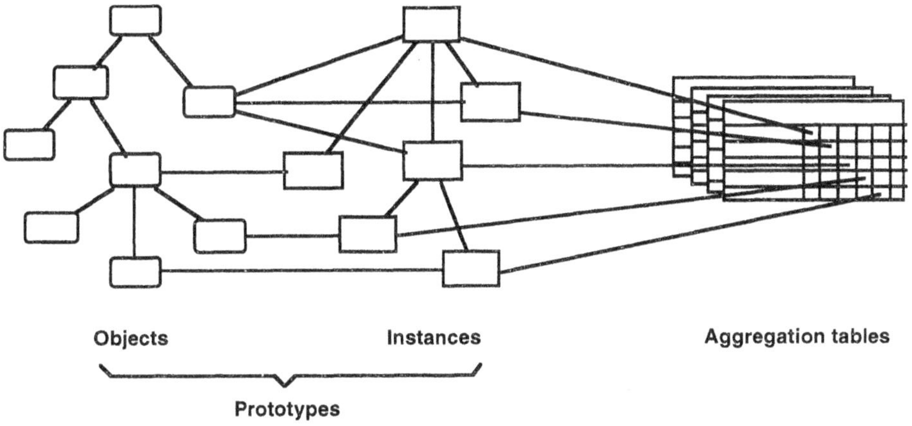

Figure 10. The schema of objects and the relationships between them.

Context thus comprises two kinds of information: settings and events. *Settings* include physical (topography and additional characteristics of the site, such as its susceptibility to earthquakes, its climate, views, etc.), cultural (built environment, customs, etc.), socio-economic, legal (codes, etc.), political, and other factors. *Events* represent the nature of the activities that will occur within and around the building. For example, in case of a restaurant, events include cooking, bringing supplies, removing garbage, parking, serving food, dining, as well as hosting birthday parties, dancing, or political rallies.

Some of the information included in the context might be translated into functional requirements, and be represented as design goals and constraints. For example, construction in an earthquake zone carries many UBC regulations and design constraints. Likewise, designing a hospital is subject to a long list of very specific requirements. However, many contextual facts are too subtle to be formalized into explicit goals and constraints. Architects are, nonetheless, aware of them, and do respond to them in their buildings. The result is what we call a 'good' building, which 'fits' within its context. Frank Lloyd Wright's Fallingwater in Bear Run, Pennsylvania, is a prime example of contextual awareness, which cannot be formalized through goals, but has nevertheless strongly influenced the design of the house.

By its very nature, *context* is inherently difficult to represent explicitly. Architects understand context by visiting the site, photographing it, rendering it in their sketchbooks, interviewing its residents, studying historical records, and, in general, spending much time there. We have proposed, therefore, an indirect representation of context, through the goals, the solutions, and primarily through the evaluators. Goals and evaluators, in particular, provide a convenient means for representing context-specific trade-off of needs, and for including problem-specific predictors and interpreters of performance. Still, we suggest that an explicit representation of context is desired, and ought to be the subject of further research. It can be achieved in the form of scripts, narratives, case studies, photo albums, and other media that have the ability to capture the above-mentioned attributes and qualities, as well as by the knowledge the architect himself brings to the design process.

5.4. EVALUATION

We consider *evaluation* to be a process that compares what *has been* achieved (or is *projected* to have been achieved) to what *ought to be* achieved. Evaluation, therefore, can be defined as *measuring the fit between achieved or expected performances and stated objectives, within a given context*. The process of evaluation can, however, only be applied to a given, specific set of performance characteristics (form, function and context), such as the form, composition and location of a building within a given site, and intended to meet the needs of a specific client, much like medical diagnosis can only be applied to the physical condition of a particular patient, under particular circumstances. When evaluating hypothetical design solutions where performances are not yet in evidence and cannot, therefore, be assessed directly, evaluation must be preceded by *prediction*. Prediction is the process whereby the expected performance of buildings (or other artifacts) is simulated, hypothesized, imagined, or otherwise made evident, so that it can be subjected to evaluation. For example, the rate of heat loss through a given building envelope must be predicted, often by way of simulation, before an evaluative procedure can determine whether this rate is acceptable for the activities that will take place in the building at a particular location. Likewise, the fire resistance properties of a wall or a door must be determined before its behavior under emergency conditions can be evaluated. Some building behaviors can be predicted by using established scientific methods, based on first principles. These include energy, structural, lighting, and other physical phenomena. Other behaviors lack such a scientifically rigorous base, and depend upon experience, rules of thumb, and sometimes sheer guesswork. These include color schemes, building *parties*, proportions, and other psychological and behavioral phenomena. Evaluation and prediction are, therefore, often value-based and dependent upon judgment, taste, and other subjective variables.

Such variables depend not only on the attributes of the solution itself, but also on the *context* in which it is embedded. The context-specificity of evaluation and prediction are, in fact, their most valuable characteristic for the development of a performance-based design paradigm. However, they are also the least computable aspect of the proposed implementation of the paradigm. Context, in the general sense that was discussed earlier, is hard to represent computationally and, therefore, difficult to incorporate in the evaluation procedures.

Moreover, evaluation refers not only the general suitability of the project as a whole to the stated goals and its context, but also to the suitability of the developing solution to goals and contexts that are particular to specific *phases* of the design, construction, and use of the building, and the concurrent relationships and influences of certain criteria on other criteria. For example, the disruption caused by the construction of a building (or a freeway) to its neighborhood may outweigh its benefits once completed. Likewise, the materials from which the building is made may be harmful to the environment or to the inhabitants of the building (asbestos and lead are prime examples).

Thus we distinguish between *Multi-Criteria* evaluation and *Multi-Phased* evaluation, where Multi-Criteria is an evaluation modality that examines a given design solution from *several different* points of view (e.g., energy, cost, structural stability, etc.), and Multi-Phased evaluation is a modality that examines how the design solution, or a succession of design solutions, satisfy a *particular design objective* (e.g., energy) throughout the study period (typically, the life-span of the building). To complicate matters even further, the designer must often engage in *both* evaluation modalities at the same time. Each modality informs the other, as well as the process as a whole. It is, unfortunately, very difficult to develop computational tools that can perform both kinds of evaluation, and at the same time, be cognizant of the particular context of the design problem. Most, if not all of the evaluation programs developed to date have chosen one of the two modalities (Kalay, 1992).

Nonetheless, as a methodology, we suggest that such multi-modal evaluation tools ought to be considered (Shaviv and Kalay, 1992).

6. Case Study

We developed a system called KAAD (Knowledge-based Assistant for Architectural Design), which is intended, among other things, to be a proof-of-concept for the proposed design paradigm and its implementation model. KAAD was designed to help architects specify design objectives, adapt existing or create new design solutions, predict and evaluate their expected performance, and compare them with the stated objectives within the specific context of designing health-care facilities for treating infectious diseases in Italy. KAAD is founded on a knowledge-base, comprising prototypical design solutions, which includes much of the information pertinent to generic building objects, such as walls, windows, doors and fixtures, as well as information specific to nursing units, infectious diseases suites, and other hospital-related data. The knowledge-base includes not only *syntactical* information (form, materials, cost, etc.), but also *semantic* information which explains the meaning and contextual relevance of the information (e.g., that a particular door in the proposed design violates infection-containment protocols). Particular solutions are derived from prototypes by adaptation to the specific context of the problem.[5]

To demonstrate the concepts discussed in this paper, we will consider the design of a small nursing unit intended to house two patients in the infectious diseases suite of a hospital located in northern Italy. A typical solution to the problem is depicted in Figure 11, which shows a collection of spaces superimposed with some access-control constraints: the gray arrows represent physical accessibility, while the white arrow represents visual access, but not a passage. For the sake of simplicity, the adjacencies with the outside and with other units of the hospital have not been represented here.

This case study, albeit of reduced complexity, is significant because hospital suites for treating infectious diseases must respond to a considerable number of constraints, many of which are often crucial. These include support for specific treatments of symptomatic seropositive or AIDS-infected patients, as well as the guarantee of adequate protection of the patients against the risk of crossed or opportunistic infections. At the same time the designer must also guarantee an adequate level of protection to the visitors and the staff by carefully considering paths, entrances, filter and reclamation areas and dressing rooms. Some for the stated goals of the design of such a nursing unit are:

---

[5] A full description of KAAD is beyond the scope of this paper, and is irrelevant for its purpose. The interested reader is referred to (Carrara et al 1994) for more details.

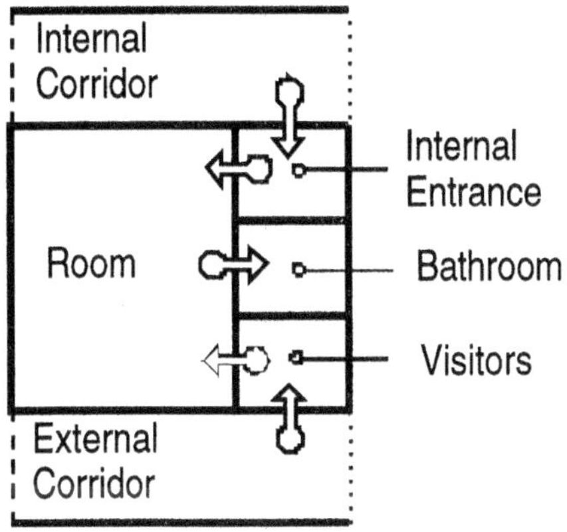

Figure 11. Layout of a typical nursing unit for treating infectious diseases.

- Proper connectivity to other hospital units
- Size limitations
- Environmental constraints, such as the number of air exchanges, air velocity, air purity, relative humidity, etc.
- Specific fixtures and furniture
- Social and personal welfare requirements.

Additionally, such suites are context-specific in term of medical procedures, and culture-dependent in terms of care and visiting patterns. The particular contextual setting of the case study included the following data:

- Location—Northern Italy, not far from the Adriatic sea, implying certain climatic and soil conditions, as well as a particular socio-economical profile of the patients and their visitors
- Date of construction/use—first decade of the 21 century (which implies certain medical procedures)
- Legislative context—an infectious diseases hospital is subject to certain rules and regulations that must be observed, in addition to customary construction methods and practices.

The event-contextual data included the following:

- Patient visiting habits, implying certain amenities for non-care-givers/receivers (e.g., parking lots, lounges, cafeteria, etc.).
- Treatment methods, including contact between staff and patients, with all the attendant risks (to both parties) that are incurred from treating patients in an infectious diseases hospital suite
- Emergency egress procedures for patients, staff, and visitors.

6.1. REPRESENTATION

For the sake of simplicity (as well as other constraints, such as time and human resources), we have chosen to combine the solutions, the goals, and the context within a single frame-based representation. Furthermore, some of the evaluators have also been represented through the same formalism, by means of procedural attachments. Two kinds of generic frames have been developed:

- *Space Units* (SU), representing classes of objects that meet requirements associated with individual rooms (or their equivalents) in a hospital, such as dimensions, type of use, environmental conditions, and so on.
- *Functional Elements* (FE), representing the physical components of the building that delimit spaces and define safety and environmental comfort levels.

These entities form two hierarchical structures: Building Units (BU), which represent the spaces in the building, and Functional Systems (FS), which represent the structure of the building.

As an example, consider the implementation of a simple SU (Space Unit) prototype of a nursing room in a hospital (Figure 12). The frame includes slots that establish the spatial relationship between this particular SU and other SUs; slots that define the maximum and minimum values for certain variables; as well as slots that define the current values associated with different variables.

6.2. PROCESS

From the user's point of view, the design process in KAAD is similar to other design processes using CAD. The user may begin by drawing lines, representing walls, doors, and other building elements, or he can begin by specifying the desired goals (constraints) the building ought to achieve, in the form of bubbles representing the functional units (rooms, corridors, outdoor spaces, etc.) and the adjacencies between them. Such drafting and modeling activities are supported by KAAD in a manner similar to other CAD systems.

The bulk of KAAD's action happens in the background. Each action of the user causes KAAD to do one of three actions:

1. search its knowledge-base for a prototype FE (Functional Element) or SU (Space Unit) which matches the characteristics of the object specified by the user, and instantiate it;
2. check to see if the modification made by the user conforms to the constraints associated with the affected FEs and SUs, and report back to the user if it does not;

3. initiate a prediction/evaluation process that will provide the user with information concerning the designed artifact as a whole (e.g., energy performance, emergency egress, etc.).

(dg3 (ako value su))	
(description (value "space unit for patient's nursing room"))	
(com (value conn1))	*Communication (part of the routes between SUs)*
(adj (value dg6 dg7))	*This SU must be adjacent to the specified SUs*
(far (value dg2 dg11 dg15 conn2))	*This SU must be far from the specified SUs*
(ims (value hfur3 ite))	*SU is an instance of specified prototype*
(sup (min 22)	*Boundary values for surface areas*
(unit mq) (max 28) (unit mq) (description "minimum and maximum net area")	
(wtemp (range 19 21)	*Range of acceptable interior winter temperatures*
(unit °C) (description "interior winter temperature")	
(stemp (range 25 27)	*Range of acceptable interior summer temperatures*
(unit °C) (description "interior summer temperature")	
(rewh (range 40 60)	*Range of acceptable winter relative humidity*
(unit %) (description "winter relative humidity")	
(resh (range 40 60)	*Range of acceptable summer relative humidity*
(unit %) (description "summer relative humidity")	
(vent (value 2)	*Desired number of air changes*
(vela (value 0.2) (unit m/sec) (description "air velocity")	*Desired air velocity*
(pura (value 4)	*Desired level or air purity*

Figure 12. An example Space Unit (SU) of a nursing unit in a hospital.

The first two actions (instantiation and checking compliance with the constraints) are transparent to the designer, triggered by the expression of goals (constraints) or the specification of design solutions. Typically, not all the information needed to completely define an FE or SU instance is given at once. KAAD follows a progressive slot-filling process, using the values specified by the designer as they become available. It automatically calculates many of the values needed to complete the instantiation, using default geometrical and material information, such as adjacencies, paths, areas, costs, and so on. Since FEs and SUs represent both goals and solutions, the two actions are conveniently similar from a programming point of view (which is, of course, why we chose to combine them in the first place).

When KAAD detects a conflict between the input and the stated constraints, it notifies the user by opening a 'warning' window (Figure 13). The user has the option to modify the input, override the constraint temporarily, or modify it (the user, in this case, is assumed to be the 'expert').

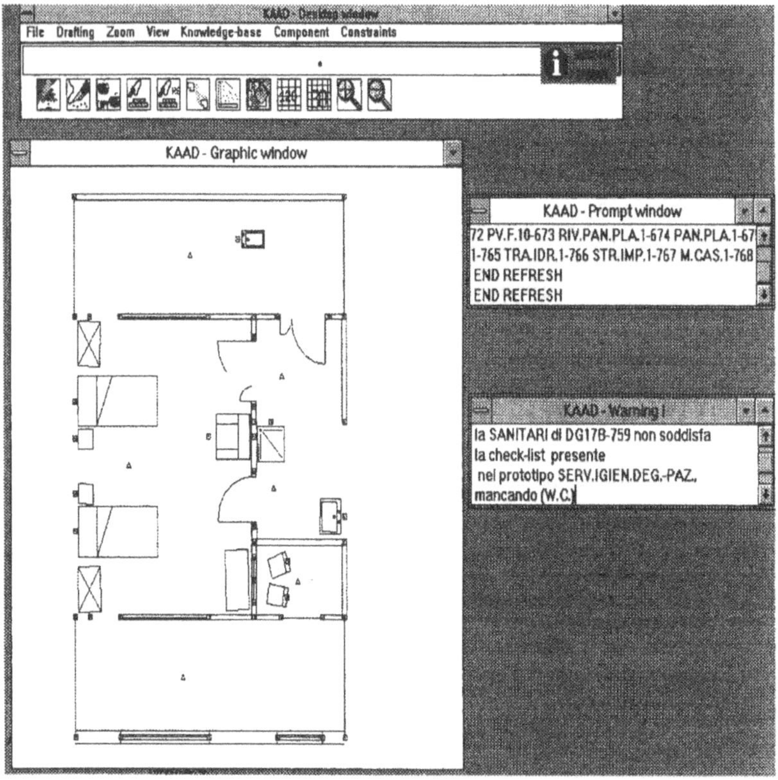

Figure 13. A typical screen of KAAD, showing a warning window.

Similarly, the user can override any number of KAAD's automatic options (including the ones normally hidden from the user), defer them, and otherwise control the actions of the system. Additionally, KAAD supports many typical CAD functions, such as grids, snapping, 2D and 3D views, automatic handling of drawing details such as insertion of doors and windows, two-, three-, and four-way wall intersections, and a full range of geometric modeling operations. Nonetheless, every *form* that the user can see on the screen (other than drafting aids), is the graphical representation of an *instance* residing in KAAD's database, along with its full *functional* description.[6]

KAAD has been implemented by a team of 12 programmers in Italy and in the USA over a five year period. They were supported by a $1.25 million grant from CNR (Italy's national research council). The knowledge-based parts of the system were implemented in Allegro Common Lisp 3.0 from Franz, Inc. The graphical and the database management components of KAAD were implemented in C. The user interface was developed under X11R3. The first prototype of KAAD was developed in the UNIX operating environment, on MicroVax and Tektronix workstations. Several PC versions, developed under Microsoft Windows 3.1 using Allegro CLl/PC 1.0 and Borland C^{++}, are in final stages of development.

7. Conclusion

The development of computational tools that can truly assist humans in performing complex activities such as architectural design relies upon developing a deep understanding of the process that is to be assisted, and on casting this understanding into a model that can be represented explicitly (and thus can be translated into a computer program). Having identified the two main characteristics of architecture as Form and Function, the search for formal theories that can explain the process of design tended to converge on causality-based paradigms. Hence the attraction of statements such as 'Form follows Function,' and its converse 'Function follows Form.' These statements provided a convenient logical foundation for design theories, much like other causalities have formed the foundation of many engineering and practically all scientific paradigms.

Many architects found these logically-convenient statements inadequate to describe what their experiences taught them. These experiences were often characterized by a discontinuity in the relationship between form and function, which they called 'the intuitive leap' (Norman, 1987). This leap occurs when architects, engaged in the search for a form that will facilitate some desired function, actually find the 'right' form. The paradigm presented in this paper attempts to recognize this experience, and use it as a

[6] As a concept-demonstration program, KAAD lacks many features that would be desirable in a more fully developed program. Particularly, it lacks means to easily extend and update its knowledge base. Changing or extending KAAD's knowledge base currently requires extensive knowledge of programming in LISP, as well as knowledge of KAAD's specific data structure. Since we have not intended KAAD to be a production CAD system, these limitations are not considered unacceptable by its design team.

basis for an alternative formal model of the design process, which can be implemented by computational means. It did not attempt to formalize the intuitive leap itself, only to accommodate it in the model. This accommodation takes the form of contextuallity: the convergence of form and function in a particular context. It strives to eliminate the precedence of either form or function, and hence of the causal relationship between the two. To compensate for this, it develops the notion of *context-based performance*, as a means for interpreting and determining the confluence of the two entities.

The novelty of the proposed approach lies in considering form and function equal, hence deserving explicit representation when implemented in computational design systems, and in striving to explicitly represent the context of the design project. The particular implementation described in this paper chose to represent form, function and context in bundles called Space Units (SUs) and Functional Systems (FSs), following object-oriented, frame-based programming practices. It stands to reason, however, that a more radical separation is conceivable, where the three primary entities will be represented by entirely different means. In that case, the interpretive mechanism will have to be separated as well.

The proposed design paradigm fits well with our view that computers ought to be *partners* in the design process, tools the designer can draw upon when developing forms, specifying functions, and interpreting their confluence (Swerdloff and Kalay, 1987). The partnership approach is intended to *facilitate* design but not to fully automate it. It is based on the observation that designers are able to cope with and manage complex design processes, and have, for centuries, achieved outstanding results doing so without the aid of computers. It also eliminates the immediate need to deal with difficult and (so far) intractable computational problems such as representing the processes of *learning*, *creativity*, and *judgment* as overt knowledge structures. Rather, the partnership approach, combined with the performance-based paradigm of design, permits the designer to provide these hallmarks of architectural design himself, while drawing upon those aspects of the design process that have already been successfully computed, such as a host of analyses, visual presentations, and even certain solution-generating algorithms.

8. References

Akin, O.: 1978, How do architects design, *in* J. Latombe (ed.), *Artificial Intelligence and Pattern Recognition in Computer-Aided Design*, IFIP, North-Holland, New York, NY.

Aksoylu, Y.: 1982, Two different systematic approaches to design, *Technical Report*, University of California, Berkeley, CA.

Alexander, C.: 1964, *Notes on the Synthesis of Form*, Harvard University Press, Cambridge, Massachusetts.
Alexander, C., Ishikawa, S., Silverstein, M., Jacobson, M., Fiksdahl-King, I. and Angel, S.: 1977, *A Pattern Language*, Oxford University Press, Oxford.
van Andel, J.: 1988, Expert systems in environmental psychology, *JAPS10 Conference*, Delft, The Netherlands.
Archea, J.: 1987, Puzzle-making: What architects do when no one is looking, *in* Y. E. Kalay (ed.), *Computability of Design*, Wiley Interscience, New York.
Armour, G. C. and Buffa, E. S.: 1968, A heuristic algorithm and simulation approach to relative location of facilities, *Management Science*, **9**(2), 294-309.
Bijl, A.: 1987, An approach to design theory, *in* H. Yoshikawa and E. Warman (eds), *Design Theory in CAD*, North-Holland, Amsterdam.
Carrara, G., Kalay, Y. E. and Novembri, G.: 1994, Knowledge-based computational support for architectural design, *Automation in Construction*, **3**(2-3), 123-142.
Coyne, R. D., Rosenman, M. A., Radford, A. D., Balachandran, M. and Gero, J. S.: 1990, *Knowledge-Based Design Systems*, Addison-Wesley, Reading, MA.
Cromley, E. C.: 1990, Sleeping around: A history of American beds and bedrooms, *Journal of Design History*, **3**(1), 1-17.
Flemming, U.: 1994, Case-based design in the SEED system, *in* G. Carrara and Y. E. Kalay (eds), *Knowledge-Based Computer-Aided Architectural Design*, Elsevier Science Publishers, Amsterdam.
Gross, M. D. and Zimring, C.: 1992, Predicting wayfinding behavior in buildings: A schema-based approach, *in* Y. E. Kalay (ed.) *Evaluating and Predicting Design Performance*, Wiley Interscience, New York.
Kalay, Y. E. (ed.): 1992, *Evaluating and Predicting Design Performance*, Wiley Interscience, New York.
Kim, M. K.: 1987, Development of machine intelligence for inference of design intent implicit in design specifications, *in* Y. E. Kalay (ed.), *Computability of Design*, Wiley Interscience, New York.
Kuhn, T.: 1962, *The Structure of Scientific Revolutions*, University of Chicago Press, Chicago.
Lenat, D. B. and Feigenbaum, E. E.: 1991, On the thresholds of knowledge, *Artificial Intelligence*, **47**(1-3), 185-250.
Laird, J., Rosenbloom, P. and Newell, A.: 1986, *Universal Subgoaling and Chunking*, Kluwer Academic Publishers, Boston, MA.
Oxman, R.: 1992, Multiple operative and interactive modes in knowledge-based design systems, *in* Y. E. Kalay (ed.), *Evaluating and Predicting Design Performance*, Wiley Interscience, New York.
Norman, R. B.: 1987, Intuitive design and computation, *in* Y. E. Kalay (ed.), *Computability of Design*, Wiley Interscience, New York.
Simon, H. A.: 1969, *The Sciences of the Artificial*, MIT Press, Cambridge, MA.
Simon, H. A.: 1975, Style in design, *in* C. Eastman (ed.), *Spatial Synthesis in Computer-Aided Design*, John Wiley, New York.
Simon, H. A.: 1979, *Models of Thought*, Yale University Press, New Haven, CT.
Shaviv, E. and Gali, D.: 1974, A model for space allocation in complex buildings, *Build International*, **7**(6), 493-518.
Shaviv, E. and Kalay, Y. E.: 1992, Combined procedural and heuristic method to energy-conscious building design and evaluation, *in* Y. E. Kalay (ed.), *Evaluating and Predicting Design Performance*, Wiley Interscience, New York.
Swerdloff, L. M. and Kalay, Y. E.: 1987, A partnership approach to computer-aided design, *in* Y. E. Kalay (ed.), *Computability of Design*, John Wiley and Sons, New York.
Venturi, R., Scott-Brown, D. and Izenour, S.: 1972, *Learning from Las Vegas*, MIT Press, Cambridge, MA.

8

A FORMAL METHOD FOR ASSESSING PRODUCT PERFORMANCE AT THE CONCEPTUAL STAGE OF THE DESIGN PROCESS

PAUL RODGERS, ALISTAIR PATTERSON AND DEREK WILSON
University of Westminster, United Kingdom

Abstract. The paper describes a formal methodology for defining and assessing product performance and its implementation in a prototype computer system. The methodology is based on high level abstract descriptions of the operations conducted within the design process. It is consequently extremely generic and succeeds in formally bridging the gap between physical product performance and actual end-user requirements. The methodology is based on defining product attributes as observable behaviour of the product in use. Defining an attribute in this way inherently reflects its required interaction with the end-user and consequently can truly be said to be in "end-user terms". A product will have a range of attributes and a performance indicator is found by combining them in a way that reflects their relative importance to the end-user. At the conceptual stage of the design process, however, the actual product does not exist, only some representation of it. To assess products at this stage requires a model or simulation of its attributes. This methodology has been implemented in a prototype Computer Aided Design Evaluation Tool (CADET) and tested with an existing product range. An example of which is presented within the paper.

1. Introduction

Human beings have always designed things. One of the most elementary traits of human beings is that they make a wide range of artifacts and tools to suit their own needs. As those needs alter, and as artifact users' reflect on the currently-available artifacts, so refinements are made to the artifacts, and completely new kinds of artifacts and tools are created and manufactured (Cross, 1994).

In the past twenty years or so there has been a significant cultural change towards manufactured goods in that product designers and manufacturers have passed through the period in which it was a challenge to manufacture an artifact to one in which the challenge is to 'Design and Manufacture' a product that satisfies user needs, wants or desires (JIDPO, 1990).

Today the quality of many products reaches such a high standard that it becomes very difficult to evaluate their intra-quality differences. Product users' judge manufactured goods not on a good-bad criterion, but on like-dislike preferences. For example Akita (1991), suggests that beauty and user-friendliness is more important than the sense of high technology within high-tech products, such as cameras, personal computers, and audio-visual equipment etc. Indeed Sipek (1993), goes as far to state that product designers have forgotten that their designed artifacts are made for people to use.

Potential users range widely, from the very young to the very old, men to women, healthy people to hospital patients, amateurs to professionals and so on. Therefore equipment should be designed to be adaptable, or in some cases specific to different peoples needs, in the most satisfying and efficient way for their personal use. This poses a new set of challenges for the designer in that a design proposal has to be evaluated at the concept stage of the design process, prior to detailed design, when s/he does not have a physical artifact, and no definite knowledge of how the market will respond to it, but simply a representation of it, for example; a design drawing or a 3-D prototype model.

This challenge presents a new requirement to the work of design in that there is a need to create a methodology to evaluate designs more accurately and earlier in the design process (conceptual stage) that ideally has some universal characteristics. It is unlikely that there is a first law of design analogous to the first law of thermodynamics, but nonetheless there is a need for a procedure with a quantifiable result to guide the designer towards his goal of satisfying the needs, wants or desires of the user.

The paper presents a review of current progress in a research project, which is based on the work of Alexander (1964), in particular, who sought to introduce a generic methodology which could satisfy the needs of designers to describe and evaluate their designs. These basic ideas have been developed into a comprehensive methodology which has been instantiated into a framework for a Computer Aided Design Evaluation Tool (CADET), described further in Rodgers et al. (1993) and Rodgers et al. (1994).

Many authors including Ulrich and Seering (1988), Miles and Moore (1989), and Hollins and Pugh (1990) highlight the neglect of research activity into the early stages of product design and manufacture, for example the concept design evaluation stage. They suggest that this may be because concept design evaluation is generally subjective in nature, relying heavily on the knowledge, intuition and experience of designers and engineers and therefore does not readily lend itself to formal expression.

Cross (1994), suggests that although there may be many different models of the design process, they all have one thing in common - the need to improve on traditional methods of working in design. There are several reasons for this interest in developing new design methodologies, strategies and procedures, including for example:

1. The fact that design problems designers have to solve nowadays has become extremely complex, for example industrial and plant machinery. Demands concerning materials and manufacturing processes' information, for example, is now so vast that it is well beyond the grasp of the individual designer to keep up to date.
2. Costs and investments involved in design projects are now so great, for example in setting-up of plant and machinery, purchase of raw materials, etc. that there are now greater pressures on the designer or design team to get it right first time before the project goes into production. Table 1 illustrates the costs involved throughout the various stages of the design of a new product.

Table 1: Costs involved in New Product Design (Hollins and Pugh, 1990).

Market research	6.9%
Product design specification	5.5%
Concept design	12%
Detail design	17.5%
Manufacturing	45.7%
Selling	12.5%

3. The fact that the needs of end-users are perceived as having far more relevance nowadays, subsequently adding to the demands placed on designers (Heskett, 1992).

2. Alexander's Design Model

The presented method of product performance assessment is based on a formal model of the design process developed from that of Alexander (1964). Alexander developed his methodology in an attempt to help designers solve increasingly complex problems. Alexander highlighted the fact that the information required to solve even the simplest design problem is well beyond the limits of the individual designer. In an attempt to rectify this shortcoming designers tend more and more nowadays to rely on their intuition, personal experiences, gut-feeling and limited knowledge when making decisions throughout the various stages of product design. For example choosing materials, deciding product finishes, manufacturing methods to be utilised and so on. Alexander's main argument is that by relying on judgment and intuition alone, what is at best vague, but more often wholly inadequate. He asserts that by adopting logical structures to the process of design, this will result in making clear or explicit what is exactly involved or required in the design process.

Alexander's design model is based on reducing the gap that exists between the designer and the user. The core of Alexander's approach is a formal description and representation of the design problem. The work presented here acknowledges the critiques that have been made of Alexander's work, by, for example, Lawson (1990). Lawson suggests that Alexander's work leads to a "rather mechanistic view" of design problems and illustrates this by pointing out two notions that are now commonly rejected:

1. That there exists an exhaustive set of requirements which can be listed at the start of the design process.
2. The listed requirements are all of equal value.

The work presented here directly addresses the second criticism of Alexander's method, specifically his listing of requirements being of equal value. It is fairly obvious that certain requirements of products are more important than others. This work incorporates this view by weighting the requirements and combining them, in turn, to give an overall measure of the product's performance.

In particular, this work is only utilising Alexander's analysis of the process, it is not utilising his suggested method of solution. The design model reported here is also extended to deal explicitly with mass produced manufactured items rather than 'one off' constructed items of architecture.

The design model utilises operators and entities to describe what are believed to be the fundamental actions and their objects respectively. Although the associated diagrams give the appearance of a procedural or flowchart model, they are not. They are intended to illustrate the abstract functional relationships between the various operators[1].

2.1. UNSELFCONSCIOUS PROCESS

The ultimate objective of design is form. Every design problem commences with an attempt to attain 'fitness' between the two entities of: (i) the form, and (ii) its context. The form is the solution to the design problem; the context determines the design problem. Usually, in product design the actual goal for the designer is not the form alone, but the ensemble comprising the form and its context. 'Good fit' is a required characteristic of the ensemble. For example a kettle (*form*) should be able to rest safely on a work-surface (*context*) within a kitchen (*ensemble*) (Alexander, 1964: 15).

Alexander (1964: 48) describes the unselfconscious process as a type of built-in fixity—types of myth, tradition and taboo which oppose strong modification. Creators of form will only introduce modification under

---

[1] An abstract function defines the relationship between two sets independent of the actual set membership (Blyth 1975).

sound compulsion where there are strong and obvious errors ("misfits") within the existing forms which demand correction.

In the unselfconscious process the designer operates directly on an actual form within its actual context. Determining the fit of form in context is by direct observation of the ensemble and in the absence of fit the design process determines actions to eliminate the misfits. An example of the unselfconscious process is the bespoke tailor fitting a suit to a client. The designer (bespoke tailor) will observe misfits in the ensemble of form (suit) and context (client) and make changes such as letting in or taking out seams. Formally the process is illustrated in Figure 1.

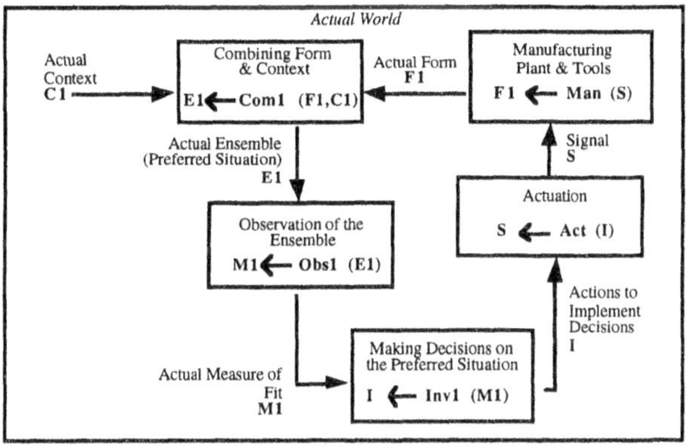

Figure 1. The unselfconscious process.

The nature of this abstract approach is that the work is not, at this stage, defining the content of entities or operators, but only the functional relationships between them. Figure 1 is explicitly **not** intended as a flowchart representation of the design process.

Actual Form **F1** and actual context **C1** are (real) entities combined by the operator **Com1** to create the ensemble entity **E1**. The designer observes the ensemble by operator **Obs1** to determine the misfit entities **M1**. The designer then applies some cognitive process described by the operator **Inv1** to determine the entity **I** of actions to be taken to eliminate the misfits. These actions are realised in the ensemble by actual tools or plant.

The actions are information in the memory of the designer which have to be converted into real physical events. This is done by a process of actuation described by the operator **Act** which converts information into power denoted by the entity **S**. The entity **S** has both physical and informational significance in that it must be sufficient to cause an effect but controlled to produce the correct effect. For example the operator **Act** may describe an actuator such as a servo motor which (via other hardware) drives a machine tool from instructions **I** in the form of a part program. The output from the servo motor must supply sufficient power to cut material but with sufficient accuracy to cut it in the way required. The effect of **S** in producing a new form **F1** depends on the actual machinery it connects to which is in turn described by the operator **Man**. These operators are intended to be complete and generic describing all processes required to generate the new form **F1**.

A more realistic model may be to describe **Man** as a differential operator, i.e. causing a change to an existing actual form rather than generating a completely new form. The given definition is used in the interests of simplicity later.

In the unselfconscious process actuation **Act** is predominately performed by the human mind in determining how tools should be operated or utilised. The unselfconscious designer need not be able to invent forms at all just respond to misfits (Alexander, 1964: 58). Most importantly the iterative modification of form to fit context occurs physically and is defined by the actual experience and satisfaction of the designer. Although an apparently obscure name the "Unselfconscious Process" is particularly apt. The designer is conscious of the form in context and responds directly to direct experience of the ensemble without consciously considering the change in form required to eliminate the misfit.

The observations made by the designer are determined by the physical configuration of the ensemble, which may be inorganic or organic, and the physical, social, cultural, and/or economic laws that apply to the configuration. Whilst the unselfconscious designer need have only intuitive knowledge of them, since he deals directly with the consequences, it will be seen later that they must be codified to produce rationalised predictions.

The unselfconscious approach is clearly unsuitable for industrially mass-produced goods for the reasons cited by Jones (1980):

1. Specifying dimensions (form) in advance of manufacture makes it possible to split up the production work into separate pieces which can be made by different people. This is the 'division of labour' which is both the strength and the weakness of industrial society.
2. Initially this advantage of defining before making made possible the planning of things that were too big for a single craftsman to make on his own, for example, large ships and buildings. Only when critical dimensions have been fixed in advance can the works of many craftsmen be made to fit together.
3. The division of labour made possible by scale drawings can be used not only to increase the size of products but also to increase their rate of production. A product which a single craftsmen would take several days to make is split up into smaller standardised components that can be made simultaneously in hours or minutes by repetitive hand labour or by machine.

2.2. THE SELFCONSCIOUS PROCESS

The method of form creation in the selfconscious process is very different from that in the unselfconscious process. Modifications are no longer made upon observation of error or misfit. They are, however, only made after a process of recognition and description by the specialist involved (Alexander, 1964: 55).

In the selfconscious process (Figure 2) the designer has become removed and is remote from the user of the product and the final physical product itself. Instead of being able to directly observe the ensemble, the designer investigates, explores and researches the actual context **C1** and constructs a mental picture of it **C2**. This process is described by the operator **Exp**.

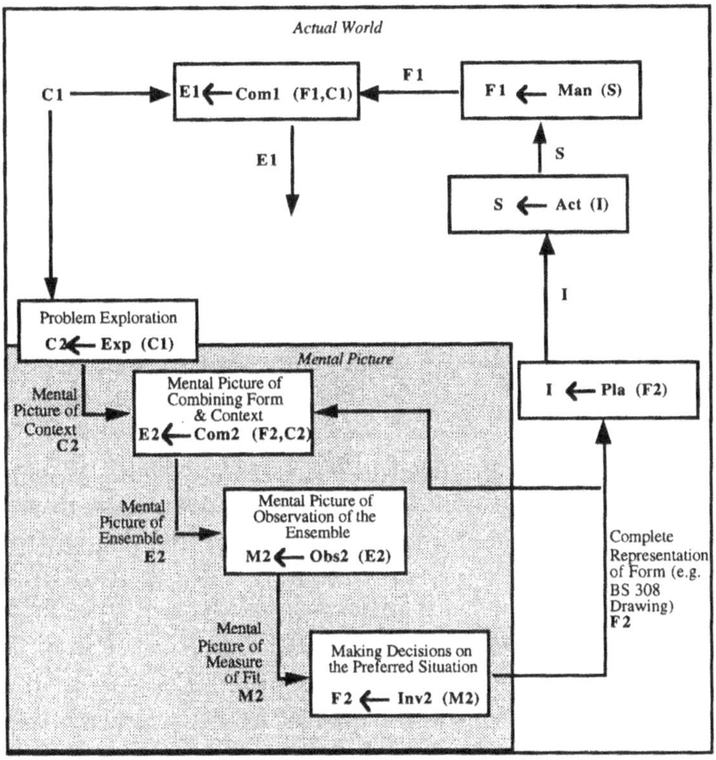

Figure 2. The selfconscious process.

Instead of working on the actual form the designer must work with a description or representation of it, **F2**, and through the operator **Com2** combine the form and context to make a mental picture of the ensemble **E2**. From this mental picture of the ensemble the designer attempts to estimate potential misfits **M2**. This process is described by the operator **Obs2**. In the same way as in the unselfconscious process the designer makes intuitive judgments, but this time on modifications to the form rather than actual actions to be taken. This process is described by the operator **Inv2**.

The process of inventing a form and physically realising it are separate. As Dormer (1993) indicates designers do not manufacture things. They think, they analyse, they may model or draw, and they specify. The most important distinction between the unselfconscious designer and the selfconscious designer is that the latter must define in detail complete and unambiguous descriptions of the shape, size, materials, and material finishes of the form prior to manufacture. Although designers may take manufacturing considerations into account there are further operations of manufacturing planning **Pla**, which determines the instructions **I** that will physically realise the form represented by **F2**, for example a BS 308 drawing (Parker, 1984).

The process of determining fit is a mental simulation of the observations the unselfconscious designer uses to determine his actions. Whilst the unselfconscious designer is dealing with reality the selfconscious designer is attempting to mentally predict a future reality.

Representation of Form

In the unselfconscious process the designer works directly with the physical form. In the selfconscious process he works on some representation of it. The representation is a complete definition of the shape, material and finish that the form will consist of[2]. When the actual form is realised measurements taken from it should correspond exactly to the measurements in the representation.

The representation of the form does not define a unique physical form. Because of inevitable tolerancing and measurement errors it defines a class of admissible actual physical forms and the actual form realised from it must be within that class.

Misfits within the Mental Picture

In the selfconscious process misfits are determined in part intuitively or by intuitively designed tests. In practice the selfconscious designer may well go through a process similar to that of the unselfconscious designer. In the example of the "off-the-peg" suit, the designer may well go through the same process of a series of successive changes or "fittings" with respect to

---

[2] The representation may contain symbolic descriptions of standard components, e.g. (electrical or electronic) but these will always be supported by shape and material representations elsewhere.

the standard mannequin as the bespoke tailor does with a client. However, unlike the bespoke tailor the result of identifying and eliminating misfits is not within the actual suit, but it is the cutting patterns representing the typical form of the suit. Although apparently identical the "off-the-peg" designer is undergoing a process of testing whilst the bespoke tailor is going through a process of production.

Difficulties with the Selfconscious Process

The operations defined within the selfconscious process are still predominately intuitive and imaginative. The designer uses drawings and diagrams to support the mental picture in his mind, however in that picture the decisions of fit of the proposed form within it are not clear. As stated initially, although the invention of form may well be intuitive, imaginative and not completely understood there is no reason why the fit of form with context should not be rationalised in an attempt to maintain the designer's intent. Alexander (1964: 77) addresses this problem by creating a formal picture of the mental picture by abstracting and defining its necessary features in formal terms. The fit of form with context can be formally defined in terms of the formal picture.

2.3. FORMAL PROCESS

Alexander asserts that within the selfconscious process the designer works entirely from the mental picture in his mind, and this picture is almost always wrong. He suggests eradicating this problem by constructing a formal picture of the design problem. This formal picture can then be scrutinised in a way not subject to the bias of language and experience (Alexander, 1964: 78). The formal picture is not intended to eliminate the intuitive and imaginative components of the design process, but to make it visible, discussable, open to criticism and make the designer accountable (Lawson, 1990).

Alexander defines the formal picture in terms of the observations of the form in context which could cause a misfit (Figure 3). The observations are called misfit variables which are either true or false. Alexander requires the selfconscious designer to state the criteria of the intuitive judgment of fit from the mental picture. Overall fit is the conjunction of the misfit variables. In the formal process the designer constructs a formal picture of the context **C3** from the mental picture of the context **C2**. This process is described here by the operator **For**. The designer combines the formal picture of form **F3** and context **C3** to produce the formal picture of the ensemble **E3**. This process is described here by the operator **Com3**. The measure or quality of fit **M3** is then determined through *predicted* observations of the ensemble **E3**. This process is described here by the operator **Obs3**. Synthesis of a new form **F3** is created in response to the predicted measure of fit or misfit **M3**. This process is described here by the operator **Inv3**. The concept representation of the form **F3** (e.g. drawing, annotated sketch, etc.) is then subject to a process of embodying the concept and adding greater detail which results in a complete and unambiguous description of the form **F2** (e.g. BS 308 Drawings). The procedure from concept of form **(F3)** to detailed description of form **(F2)** generally follows along the following lines, (French, 1985):

1. Conceptual design
2. Embodiment design
3. Detail design

This process is described here by the operator **Emb**.

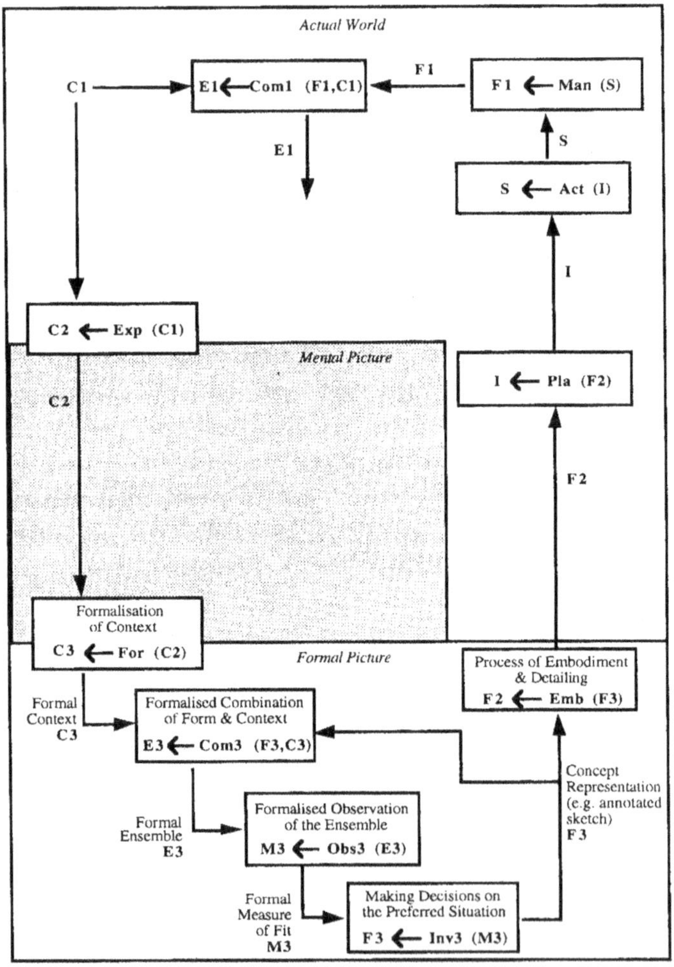

Figure 3. Formal picture of the mental picture.

Determining the formal measure of fit **M3** within the formal picture (Figure 3) is the process of product performance assessment at the conceptual stage of the design process.

2.4. ATTRIBUTES AND ENSEMBLE

Misfit Variable
Alexander's formalisation of context is based on a set of Boolean variables. However this implies that each requirement is of equal importance, a notion now widely rejected by many authors including Lawson (1990).

It is important to note that a misfit variable is defined as an observation that could be made of an actual ensemble. The paper will demonstrate later that the designer's problem at the conceptual stage of the design process is to predict this observation, since the actual ensemble does not exist.

To enable requirements of differing importance to be defined the misfit variable is extended to an attribute variable which can take on a wider range of values appropriate to the observation.

Attributes
An attribute a is defined by the set of values it may take which is defined as its type A.

$$A = \{a : S(a)\}$$

where S is an open sentence defining inclusion within A (Blyth, 1975).

$$a \in A,$$

with the constraint that A must be scalar and totally ordered.
Misfit variables (Alexander, 1964), are special cases of attribute variables equivalent to an enumerated type: {false, true}

The meaning or semantic of an attribute, for example 'consumes_fuel_efficiently', is its method of observation from an actual ensemble. An attribute observation Oba is a function from an actual ensemble **E1** to an attribute value A.

$$\text{Oba} \mid \mathbf{E1} \rightarrow A$$

An attribute is the observation of an element of performance of an actual form **F1** within an actual ensemble **E1**, i.e. the product in use. For example, a performance element of a car could be that it should consume fuel efficiently. The attribute 'consumes_fuel_efficiently', for example miles per gallon, could be directly measured under stated conditions. Another attribute such as 'looks_fast' would have to take values from an enumerated set such as {slow, average, good, quick, fast} but ultimately could only be measured from the stated responses of individuals.

It is important that attributes are based on direct observation and do not implicitly contain theories about their causes. For example a suitcase may have an attribute 'comfortable_to_carry' which could be reasonably objectively defined and evaluated in terms of muscular discomfort. It should not however be defined in terms such as weight which implicitly reflect ergonomic theories of human capacity. Such considerations are clearly essential to the assessment but are not contained within definitions of attributes.

In general attributes range from the objective to the subjective (Figure 4).

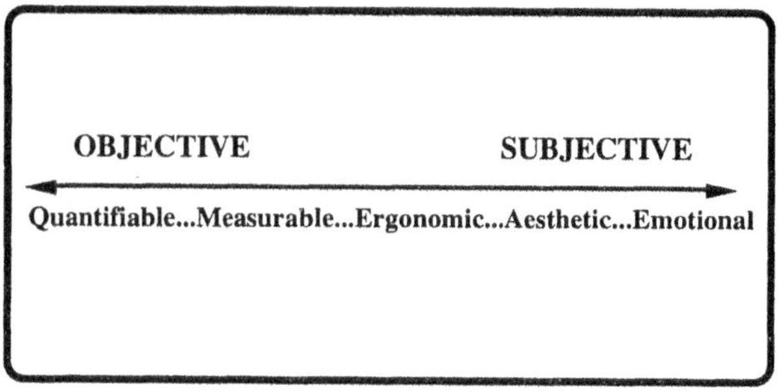

Figure 4. Attribute spectrum.

Definition of the attributes formalises the designer's view of the significant requirements of the form in use. All the attributes **A** for a form is the product set of the n individual attribute types defined,

$$\mathbf{A} = A_1 \times A_2 \times \ldots A_n.$$

The elements of **A** are the n-tuples reflecting the evaluations of each attribute,

$$\mathbf{a} \in \mathbf{A}, \quad \mathbf{a} = \langle a_1, a_2, \ldots a_n \rangle.$$

The corresponding n attribute observation functions are,

$$Oba_1, Oba_2, \ldots Oba_n,$$

and the single function which produces the attribute n-tuple is,

$$\mathbf{Oba} \mid \mathbf{E1} \rightarrow \mathbf{A}$$

$$\mathbf{a} = \langle Oba_1(\mathbf{E1}), Oba_2(\mathbf{E1}), \ldots Oba_n(\mathbf{E1}) \rangle$$

The attribute variables define the relevant aspects of the product in use.

2.5. FORM

The detailed representation of form **F2** should be a complete and unambiguous representation of an actual form **F1**. However due to the inevitable measurement and manufacturing errors, **F2** in practice defines a class of actual forms. **F2** is in fact an inclusion condition for a class of admissible actual forms **F**. The usual practice is to treat measurement as perfect and incorporate measurement error within the range of permissible actual forms. For example a component is measured and that measurement compared to a toleranced drawing. If the measurement is within tolerance then the component is, in that respect, within the class of admissible forms, otherwise it is not. It is not generally assumed that if a component is out of tolerance that it may still be within the class of admissible forms due to the measurement error.

This leads to algebraic complications since the manufacturing operator **Man** would have to be defined as producing classes of admissible actual forms rather than a single actual form. Consequently tolerancing and measurement error will be ignored and it will be assumed that there is a unique actual form **F1** associated with its representation **F2**. The relationship between **F1** and **F2** is defined by a measurement operator **Mes**,

$$\mathbf{Mes} \mid \mathbf{F1} \rightarrow \mathbf{F2}.$$

Similarly it will be assumed that a concept representation **F3** is associated with a unique actual form **F1** also related by the operator **Mes**,

$$\mathbf{Mes} \mid \mathbf{F1} \rightarrow \mathbf{F3}.$$

This of course implies that the process of embodiment, detailing **Emb**, and manufacture **Man** are completely deterministic for a given concept. This is clearly not the case in practice but makes little difference to the problem of product performance assessment at the concept stage of the design process.

2.6. PERFORMANCE

To get a single performance evaluation of the form in context the individual attributes must be combined in a way that reflects their relative importance (Lera, 1981). The final measure of performance **M3** is the formal picture of the actual performance **M1**. The combination of attributes to find **M3** is described by a combination function **Cob** which defines their relative importance.

$$\mathbf{Cob} \mid \mathbf{A} \rightarrow \mathbf{M3}.$$

This addresses Lawson's (1990) major criticism of Alexander (1964) of not accounting for the relative importance of the attributes.

2.7. CONTEXT

The formalisation of context is the combination function **Cob**. Notice that defining the combination function implicitly defines the attributes as well as explicitly defining their relative importance. This definition is equivalent to Alexander's formalisation of context when all attributes are Boolean and their combination is conjunction. Notice again that the formalisation of context is defined in terms of observations that **could** be made of an actual ensemble.

2.8. EVALUATION

The definition of product performance is illustrated below (Figure 5).

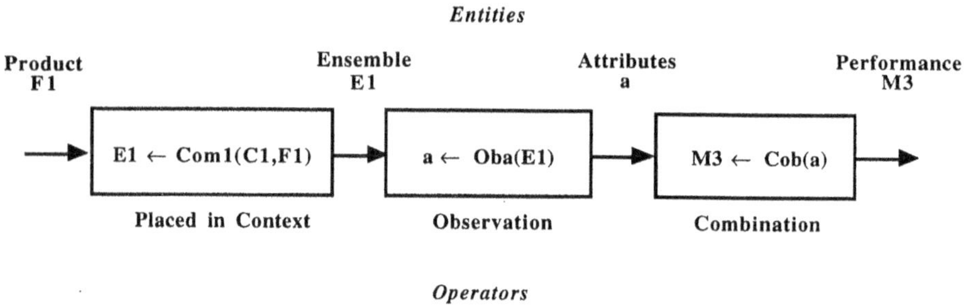

Figure 5. Performance assessment of an actual product.

The assessment of an **actual** product in context is by implementation of the operators **Oba**, which evaluates the attributes and **Cob**, which combines them into a single performance assessment. However, at the conceptual stage of the design process there is no actual product. From Figure 3, **M3** is predicted by the operator **Com3(C3,)•Obs3**. Consequently this operator is required to simulate **Oba•Cob**.

3. Product Performance Assessment

Definition of Assessment

Product performance assessment at the conceptual stage was described as implementation of operators,

Com3(C3,)•Obs3,

on the concept representation of form **F3** (see Figure 3).

The product performance assessment method has two elements of:

1. *Problem Definition*—definition of entity **C3** by implementation of operator **Exp•For**
2. *Evaluation*—implementation of operators **Com3** and **Obs3** on **F3** to produce the measure of performance **M3**.

This work will start with the evaluation problem since the requirements of evaluation influence problem definition.

At the concept stage, where there is no **actual** product, the operator **Oba** must be simulated to predict the performance **M3**. This requires prediction of each individual attribute to be combined by the operator **Cob**. The operator **Oba** is simulated using two further operators **Ext** and **Mod** to be described later.

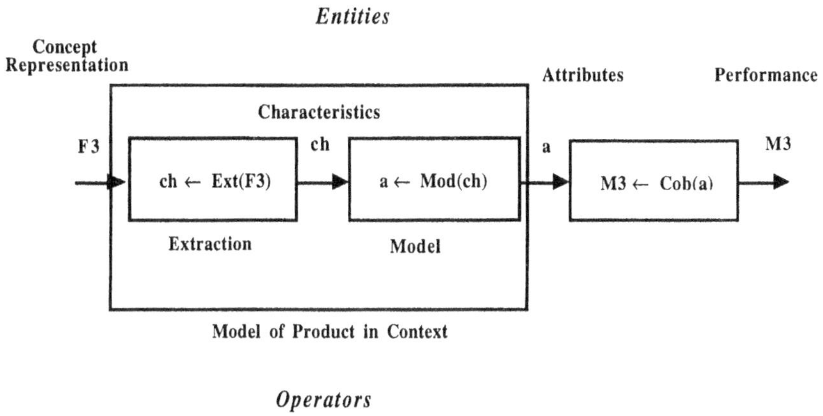

Figure 6. Performance assessment at concept representation stage.

3.1. ATTRIBUTE PREDICTION

The prediction of objective attributes such as: 'consumes_fuel_efficiently' is well defined, although in many cases may be mathematically difficult. The engineering sciences, for example thermodynamics, are predominately concerned with making these types of predictions, the results of which are embodied in available computer software, e.g. RASNA MECHANICA®.

Theoretical models for predicting behaviour are based not of the complete form itself but on specific properties extracted from it. In many cases the same property will appear in more than one attribute. It is therefore convenient and expedient to decompose the operator **Obs3** into a composition of a further operator of extraction **Ext**, a model **Mod**, and a new entity reflecting the extracted properties called characteristics **Ch**.

Characteristics are inherent properties of any product, independent of the product's use, and can be determined purely from the representation. Product characteristic examples include mass, colour, material specifications, dimensional information (length, width, height, etc.).

Characteristics have the same mathematical structure as attributes. A characteristic ch is defined by the set of values it may take which is defined as its type Ch.

$$Ch = \{ch : S(ch)\}$$

where S is an open sentence defining inclusion within Ch (Blyth, 1975).

$$ch \in Ch,$$

The characteristics of a form **Ch** are the product set of the n individual characteristic types defined for the form,

$$\mathbf{Ch} = Ch_1 \times Ch_2 \times \ldots Ch_n.$$

Consequently the elements of **Ch** are the n-tuples reflecting the determination of each characteristic,

$$\mathbf{ch} \in \mathbf{Ch}, \quad \mathbf{Ch} = \langle ch_1, ch_2, \ldots ch_n \rangle.$$

The operator **Ext** is a function from form **F3** to characteristics **Ch**.

$$\mathbf{Ext} \mid \mathbf{F3} \rightarrow \mathbf{Ch}$$

and the model **Mod** a function from characteristics **Ch** to attributes **A**.

$$\mathbf{Mod} \mid \mathbf{Ch} \rightarrow \mathbf{A}.$$

3.2. ATTRIBUTE EVALUATION EXAMPLE

The assessment method, developed within this work, is illustrated by the following example. The example will outline the assessment of a single attribute.

3.3. TOOTHBRUSH

In this example the concept representation **F3** is an annotated sketch (Figure 7). In this case the attribute is objective but less easy to define since one has had to use natural language to describe the observation rather than mathematical language representing specific observations.

Attribute

$$\text{reaches\_all\_teeth} \in \{\text{True, False}\}$$

Method of Observation

reaches_all_teeth = filament ends contact with every tooth surface in the mouth".

In this example natural language formulation has been used and is denoted within double quotation marks.

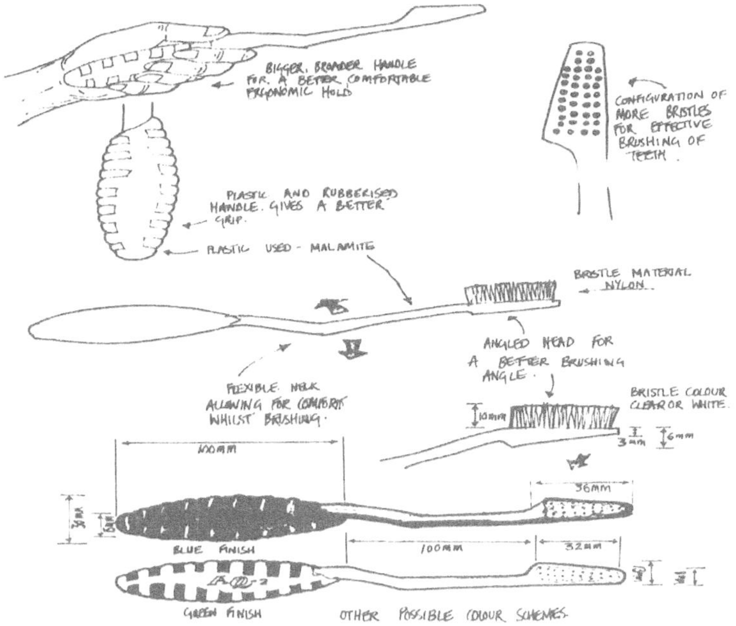

Figure 7. Design sketch of toothbrush.

Model

The model is constructed from considerations of spatial occupancy of the toothbrush drawn from computer-generated three-dimensional models representing the access region in the mouth for a given brushing action drawn from expert knowledge (Walsh and Lamb, 1992/93).

In this case the model was constructed empirically by using a plaster cast of teeth and estimates of cheek flexibility to determine the region **Access** (Figure 8). The sweep trajectory **M** is a conservative approximation to ideal brushing action determined from interviews from dentists (Walsh and Lamb, 1993). The model is expressed as :

reaches_all_teeth = ($\mathbf{SWEEP}(V,M) \cap^{*} \text{Access}^{*} = \emptyset$)

where : V is the volume describing toothbrush spatial occupancy,
M is the path it is to be moved along,
SWEEP(V,M) denotes the SWIVEL 3D™ PROFESSIONAL operator for moving a volume V along a path M,
Access* is the regularised complement of the mouth access volume,
$\cap^{*}$ denotes regularised set intersection
and Ø the empty set.

(i.e. The work requires the toothbrush to be fully contained within the Access volume for the tooth brushing action)

Characteristics

V is the volume representing the spatial occupancy of the toothbrush.

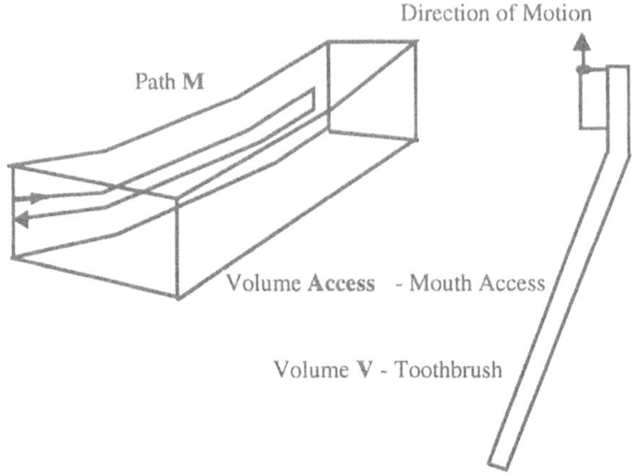

Figure 8. Diagram of access volume within mouth (empirically based model).

Attribute Evaluation

A = {<reaches_all_teeth> | reaches_all_teeth ∈ {true,false} },
Oba = "filament ends contact with every tooth surface in the mouth".
Ch = {<V> | V ∈ bounded regular subsets of **E3** },
ch = **Ext(F3)** = "Construction of V in SWIVEL 3D™ PROFESSIONAL",
Mod(ch) = "Definition of M and Access within SWIVEL 3D™ PROFESSIONAL",
a = "Execution of the **Sweep** operator in SWIVEL 3D™ PROFESSIONAL and visual inspection of the generated image (Figure 9)"
a = True

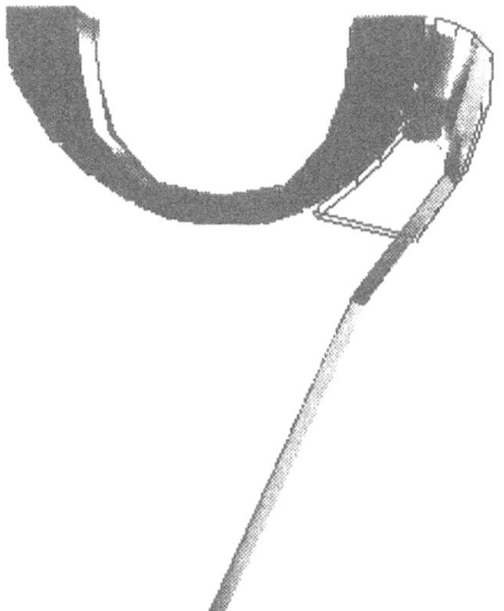

Figure 9. SWIVEL 3D™ PROFESSIONAL generated model of toothbrush in mouth.

3.4. CRITIQUE OF WORKED EXAMPLE

The example illustrates the method of assessment. The example shows the organisation of information embedded within the assessment system. The system contains the following elements:

Association between User and Designer Views

An attribute is a formalisation of the intuitive judgment that would be made by the designer, or in other words the designer's "feel" for the problem, based on the requirements of the user. However, disagreement and misunderstanding can, and does, arise between the designer and the user (of the product, system, etc.). Cross (1994), states that this is because the designer and the user focus on different aspects of the product's requirements. The user generally focuses attention on the attributes of the product and states his or her requirements in natural terms, for example 'easy_to_clean'. The designer, however, concentrates more on the product's characteristics, which seek to establish the product attributes, which in turn attempt to satisfy the users' requirements. This approach addresses the problem by formally linking the physical characteristics of the product to a clear statement of the user requirements. For example, in the case of the toothbrush, the attribute and its observation,

reaches_all_teeth = *"filament ends contact with every tooth surface in the mouth".*

clearly reflects the user requirement whilst the model links the relevant characteristics of spatial occupancy, which are under the control of the designer, to it.

Attribute Selection

It is questionable whether a complete list of requirements (attributes) can be defined for a product at the start of the design process. Many requirements of products become apparent only through the actual process of assessing design proposals.

Model

Whilst the problem of attribute selection is in determining an adequate, even if incomplete, set of attributes, the potential difficulty in the model is in accurately simulating those attributes that are defined. Moreover, the model itself, (i.e. the equation) says little to the designer about the rationale for its construction, for instance: the toothbrush example is based on ergonomics and human factors theory. If such a system is to be of use to the designer then the rationale must also be available in an explanation facility.

Attribute Type
The assessment of the attribute in the example was to predict whether the form proposed would be suitable, for example would the toothbrush 'reach_all_ teeth'. The assessment of this example was either true or false, in other words Boolean. However, it would be more useful to facilitate the designer with a numerical result that rates or scores the attributes. This would then assist the designer to address the specific characteristics that failed.

4. *FLEX* Implementation of the Evaluation Methodology

A computer implementation of the assessment method has been written in *FLEX* —an expert system toolkit that offers frame based, data driven and rule based functionality fully integrated into a PROLOG environment.

The computerisation of the assessment methodology (Figure 10) requires implementation of exactly the same operators used in the toothbrush example previously.

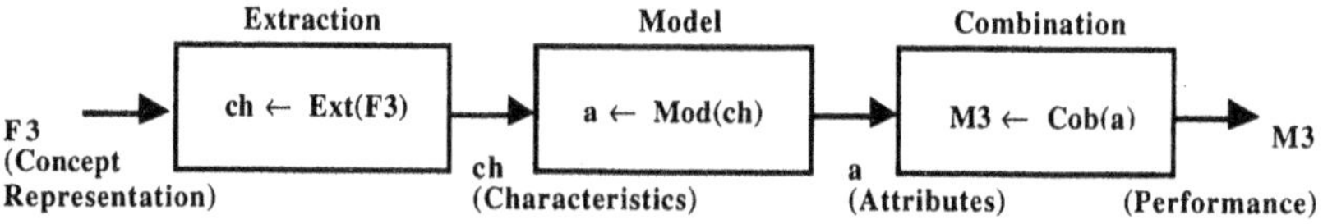

Figure 10. CADET tool implementation requirements of Obs3.

FLEX implementation of the previous defined operators, **Ext, Mod,** and **Cob** is required. The implementations are illustrated using the performance of an actual product, i.e. a toothbrush, as an example. In this example the attributes of the toothbrush and their methods of observation are:

A1: long_lasting ∈ {0..100}
Oba1: long_lasting = "lasts a minimum of three months without wearing out"

A2: comfortable_to_hold ∈ {0..100}
Oba2: comfortable_to_hold = "comfortable to hold whilst brushing teeth"

A3: removes_plaque_efficiently ∈ {0..100}
Oba3: removes_plaque_efficiently = "removes enough plaque each time to prevent a significant amount of tooth decay"

A4: does_not_irritate_gums ∈ {0..100}
Oba4: does_not_irritate_gums = "doesn't make gums bleed or cause sore gums"

A5: reaches_all_teeth ∈ {0..100}
Oba5: reaches_all_teeth = "filament ends contact with every tooth surface in the mouth"

A6: looks_attractive ∈ {0..100}
Oba6 : looks_attractive = "looks attractive within a bathroom environment"

The attribute set for this problem is:

$$\mathbf{A} = A1 \times A2 \times A3 \times A4 \times A5 \times A6,$$

a ∈ **A** = <long_lasting,comfortable_to_hold,removes_plaque_efficiently, does_not_irritate_gums,reaches_all_teeth, looks_attractive >.

and the corresponding observation set,

{"lasts a minimum of three months without wearing out", "comfortable to hold whilst brushing teeth", "removes enough plaque each time to prevent a significant amount of tooth decay", "doesn't make gums bleed or cause sore gums", "filament ends contact with every tooth surface in the mouth", "looks attractive within a bathroom environment"}.

The combination function in this case is a linear weighting of the attributes,

$$\mathbf{Cob}\,(a) = \Sigma_{1 \le i \le 6}\, w_i\, a_i.$$

4.1. *FLEX* IMPLEMENTATION OF OPERATOR **COB**

Operator **Cob** is implemented by the *FLEX* structure action. Within this structure each attribute is identified by a *FLEX* relation identifier of the same name (Figure 11).

```
combination function

action combination_function(Name,Score);

do restart
and  looks_attractive(Name,Score)
and  reaches_all_teeth(Name,Score)
and  comfortable_to_hold(Name,Score)
and  does_not_irritate_gums(Name,Score)
and  lasts_long(Name,Score)
and  removes_plaque_efficiently(Name,Score)

and total_performance(Name,Score) =
looks_attractive(S1 * Weight) + reaches_all_teeth(S2 * Weight) +
comfortable_to_hold(S3 * Weight) + does_not_irritate_gums(S4 * Weight) +
lasts_long(S5 * Weight) + removes_plaque_efficiently(S6 * Weight) / 6 * (100).
```

Figure 11. Combination function.

4.2. *FLEX* IMPLEMENTATION OF OPERATOR **MOD**

The model for each of the attributes in the combination function is coded as a relation in *FLEX* based on knowledge extracted from experts, for example Walsh and Lamb (1992/93), and Delaunay (1982).

For example the model for the attribute 'does_not_irritate_gums' is a model of the observation "doesn't make gums bleed or cause sore gums" and is shown below (Figure 12).

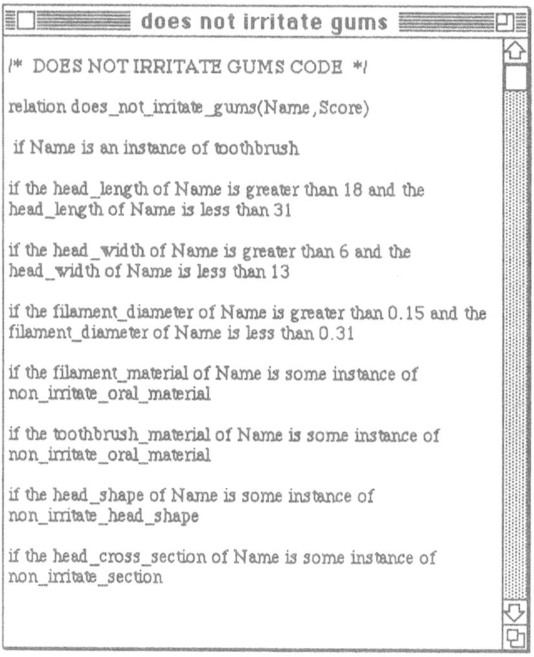

Figure 12. 'does_not_irritate_gums' model.

The model consists of a collection of clauses based on either the product characteristics or sub-relations. For example, the first clause, 'if Name is an instance of toothbrush', ties the attribute to the class of product defined as toothbrushes. The second clause, 'if the head_length of Name is greater than 18 and the head_length of Name is less than 31', is directly based on the product characteristic 'head_length' and reflects expert opinion (Chong and Beech, 1983). The fifth clause, 'if the filament_material of Name is some instance of {non_irritate_oral_material}', is based on the sub-relation '{non_irritate_oral_ material}'. This sub-relation is intended to be applicable to any item which is placed in the mouth and forms part of a library of similar sub-relations. This is implemented in *FLEX* by the following code shown below (Figure 13).

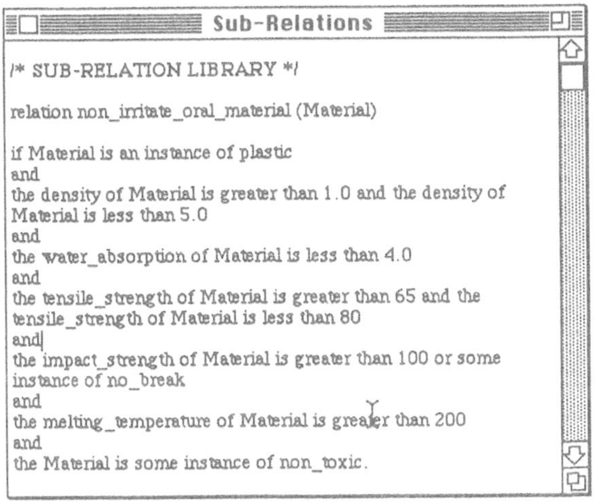

Figure 13. CADET tool library of sub-relations.

Each attribute has a similar model and their combination is the *FLEX* implementation of the operator **Mod**.

4.3. *FLEX* IMPLEMENTATION OF OPERATOR **EXT**

Each of the attribute models defines the characteristics necessary for its computation. The list of characteristics required are shown below (Figure 14). Notice that certain product characteristics such as 'filament_diameter', 'handle_cross_section' and 'head_shape' occur in more than one attribute model. The process of extraction is manual. The designer must inspect what ever concept representation s/he is assessing in response to system prompts. In principle however if the system was linked to a CAD system then the type of feature extraction facilities available on some systems could be exploited either to automatically extract the characteristics or at least provide user friendly interactive methods, analogous to the systems described by Tovey (1994) and Buck (1992/93).

5. Example of the CADET System in Use

The CADET system may be used for either a total evaluation or for individual attribute evaluations. Each attribute can now be computed by selecting it from the pull down menu (Figure 15). The designer is requested to fill in the product characteristics describing his or her concept design proposal at the CADET system dialog box prompt (Figure 16), in this case actual characteristics of the toothbrush concept design proposed, e.g. 'toothbrush_length','handle_thickness', 'handle_cross_section', etc.

Product Characteristics \ Attributes, in users' terms	LONG LASTING	COMFORTABLE TO HOLD	REMOVES PLAQUE EFFICIENTLY	DOES NOT IRRITATE GUMS	REACHES ALL TEETH	LOOKS ATTRACTIVE
Handle Length		X				
Toothbrush Length		X			X	
Head Length				X	X	
Filament Length	X		X		X	
Filament Diameter	X		X	X		
Handle Width		X				
Head Width				X		
Handle Thickness		X				
Head Thickness						
Number of Filaments in One Tuft (Packing Density)	X					
Number of Tufts in Head						
Filament Material	X		X	X		
Toothbrush Material	X			X		X
Head Shape				X	X	X
Handle Shape		X	X			X
Filament-End Shape	X					
Handle Cross-Section		X				X
Head Cross-Section				X		
Tuft Arrangement			X			
Toothbrush Colour(s)						X
Filament Colour(s)						X
Toothbrush Finish		X				
Angle between Toothbrush Head & Handle			X		X	

Figure 14. Selection of attributes with product characteristics required to construct each model.

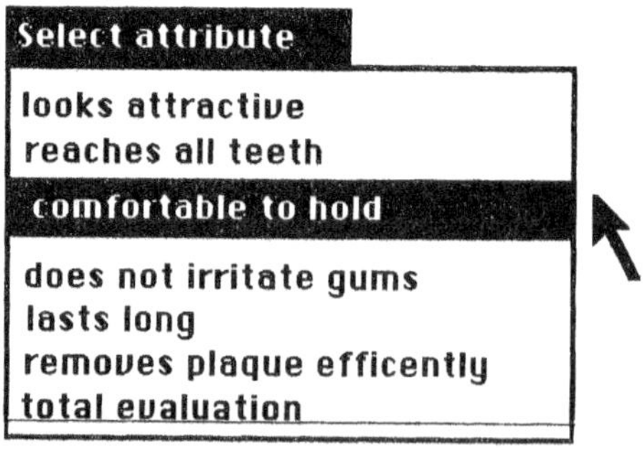

Figure 15. Attribute selection menu.

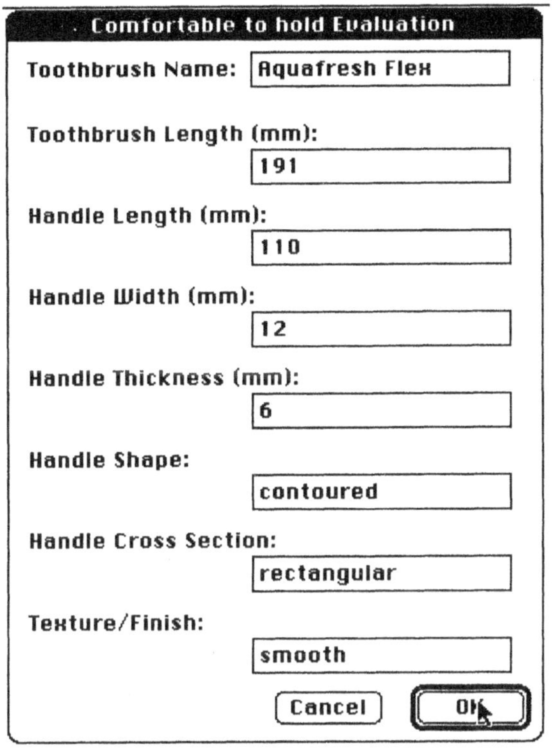

Figure 16. CADET system product characteristics extraction dialog.

Notice that the system obliges the designer to have defined sufficient detail for the concept to be evaluated. Having entered the product characteristic data into the system the designer can then quickly evaluate the potential for success of his or her concept design proposal.

The result is displayed which gives the designer a numerical indication of how well or how badly the concept proposed has done (Figure 17).

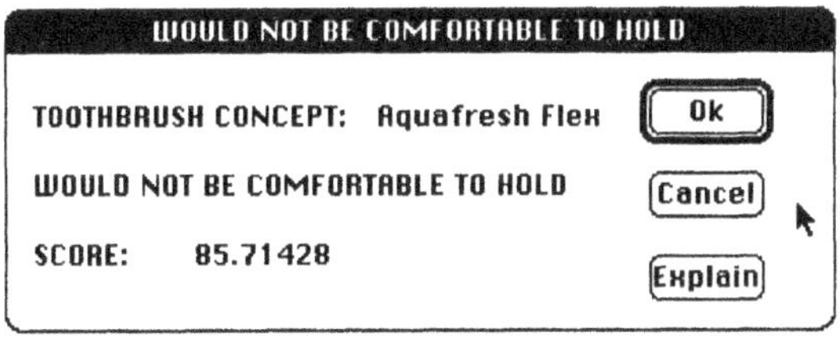

Figure 17. CADET system evaluation dialog.

The designer may investigate the reasons for the evaluation by referring back to the *FLEX* relations previously described. However as was found in the previous examples whilst the *FLEX* language makes the calculations clear the underlying rational is not apparent. To achieve this an explanation facility containing the expert knowledge used is available for each attribute (Figure 18).

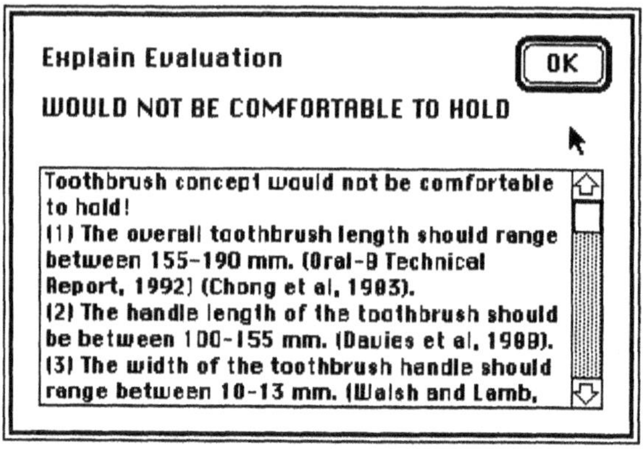

Figure 18. CADET system explanation facility.

6. Conclusions

The paper has presented a new methodology for assessing product performance in user rather than product terms. The methodology developed addresses the problem of predicting the measure of fit of product design proposals at the conceptual stage of the design process.

The method of assessment is made up of three major stages:

1. Defining the attributes, in users' terms.
2. Determining the model of each attribute.
3. Defining the combination function.

The intention for the future is to consolidate and further verify this work by introducing the following features:

1. Verification of CADET System Testing
The CADET System evaluation results will be tested against conclusions drawn from conducting appropriately controlled experiments on different groups of parties, for example product designers, manufacturers and users. These tests will seek to determine if:

(a) The evaluations are equivalent
(b) They are equivalent for the same reasons as those embodied within the system

or otherwise.

Formally in terms of the system description the tests will seek to establish,

(a) If the combination function **Cob** contains sufficient and relevant attributes

(b) If the operator **Mod** is an accurate model of the observations it is intended to predict.

2. Assessment of Other Products

A great deal of design work in practice is concerned not with the creation of radical new design concepts but with the making of modifications to existing product designs. These modifications seek to improve a product—to improve its performance, to enhance its appearance, and so on.

The prototype CADET System is at present being tested against established product designs, particularly consumer products whose purpose and use is well defined, for example: telephones, electric kettles, electric hair-driers, system and disposable shavers. This type of product has evolved through several generations of products as end-user needs have developed.

3. Design Model Development

For the designer the CADET assessment of product performance is an optimisation criteria s/he is working towards, i.e. s/he is attempting to maximise the rated performance of the concept. The design model developed so far has deliberately not had a chronological element, it has only demonstrated the relationship between defined operations. It is intended to develop this model into a chronological model by treating it as an optimisation problem of performance. It is also anticipated that the design process will be described by some form of genetic algorithm within this.

7. References

Akita, M.: 1991, Design and ergonomics, *Ergonomics*, **34**(6), 815-824

Alexander, C.: 1964, *Notes on the Synthesis of Form*, Harvard University Press, Cambridge, Massachusetts.

Blyth, T.S.: 1975, *Set Theory and Abstract Algebra*, Longman, London.

Buck, P.: 1992-93, Providing intelligent support for design, *Manufacturing Intelligence*, **Winter**, 10-13.

Chong, M. P. and Beech, D. R.: 1983, Characteristics of toothbrushes, *Australian Dental Journal*, **28**(4), 202-211

Cross, N.: 1994, *Engineering Design Methods: Strategies for Product Design* (2nd edn), John Wiley, Chichester.

Delaunay, P.: 1982, Some current data involved in the choice of toothbrush, *Actualites Odonto-Stomatologiques*, **138**, 249-258.

Dormer, P.: 1993, *Design Since 1945*, Thames and Hudson, London.

French, M. J.: 1985, *Conceptual Design for Engineers* (2nd edn), The Design Council, London.

Heskett, J.: 1992, Product integrity, *Innovation*, **Spring**, 17-19.

Hollins, B. and Pugh, S.: 1990, *Successful Product Design: What to do and when*, Butterworth, London.

Japan Industrial Design Promotion Organization (ed.): 1990, *Good Design Products 1990*, JIDPO, Japan.

Jones, J. C.: 1980 edn, *Design Methods: Seeds of human futures*, John Wiley, Chichester.

Lawson, B.: 1990, *How Designers Think—The Design Process Demystified* (2nd edn), Butterworth Architecture, London.

Lera, S. G.: 1981, Architectural designers' values and the evaluation of their designs, *Design Studies*, **2**(3), 131-137.

Miles, J. C. and Moore, C. J.: 1989, An expert system for the conceptual design of bridges, *Proceedings of the Artificial Intelligence in Civil and Structural Engineering Conference*, pp. 171-176.

Parker, M. (ed.): 1984, *Manual of British Standards in Engineering Drawing and Design*, British Standards Institution/Hutchinson, London.

Rodgers, P. A., Patterson, A. C. and Wilson, D. R.: 1993, A computer-aided evaluation system for assessing design concepts, *Manufacturing Intelligence*, **15**, 15-17.

Rodgers, P. A., Patterson, A. C. and Wilson, D. R.: 1994, Evaluating the relationship between product and user, *IEE Computing and Control Division Colloquium on Customer Driven Quality in Product Design*, Digest No: 1994/086.

Sipek, B. (ed.): 1993, *The International Design Yearbook*, Laurence King, New York.

Tovey, M.: 1994, Form creation techniques for automotive CAD, *Design Studies*, **15**(1), 85-114.

Ulrich, K. and Seering, W.: 1988, Computation and conceptual design, *Robotics and Computer-Integrated Manufacturing*, **4**(3/4), 309-315.

Walsh, T. F. and Lamb, D. J.: 1992/93, Update of oral hygiene aids: Toothbrushes, *Dental Health*, **31**(6), 3-5.

Walsh, T. F. and Lamb, D. J.: 1993, Research visit to conduct knowledge engineering exercise with T. F. Walsh and D. J. Lamb, Department of Restorative Dentistry, School of Clinical Dentistry, Sheffield University Dental School, 19 March.

9

ON A SYMBOLIC CAD FRONT-END FOR DESIGN EVALUATION BASED ON THE PI-THEOREM

STEPHAN RUDOLPH
Stuttgart University, Germany

Abstract. The current implementation and theoretical foundation of a possible symbolic front-end for CAD systems for the evaluation of engineering design objects during the design process is described. Based on implicit functional descriptions of the design object (i.e. the design parameters contained in the database of the solid modeler), the Pi-Theorem is used to derive the associated dimensionless groups. Based on the assumed validity of the evaluation hypothesis that *"any minimal description in the sense of the Pi-Theorem is an evaluation"*, these automatically generated dimensionless groups serve then as a symbolic representation for the purpose of design object evaluation.

1. Introduction

Due to the complexity of problems inherent to the design process of engineering objects, there has been a significant amount of effort to support the designer by means of CAD/CAE systems able to ease many of the designers routine tasks (SDRC, 1993). Since the benefit and gain of productivity of such software systems is widely accepted, the development of software systems with an even greater functionality is an area of intensive current research (ten Hagen et al., 1991; Dym, 1994).

While in many new developments the emphasis lies on the application of new AI-based techniques to the area of design, there has also been a significant effort in the traditional engineering community to formalize the design process from an engineering viewpoint (Pahl and Beitz, 1993; Suh, 1990). Since the design of a technical product involves the definition of its purpose, functional descriptions defined as functional relationships between physical input, output and state variables may be used throughout the design process to represent design object properties independent of a particular solution (VDI, 1987). During the design process, this functional description becomes then more and more concrete (Andreasen, 1992).

Since the formalism of dimensional analysis based on the Pi-Theorem requires only qualitative information about the relevance list of the physical design parameters, this method is most ideally suited for processing qualitative physical knowledge encoded in such functional descriptions of the design object. Dimensional analysis has therefore been already been applied to engineering design problems in the past (Kloberdanz, 1991; Dolinskii, 1990), where it was used to ease the modeling and helped to gain a deeper understanding of the functional behavior of the design object. Other works using dimensional analysis as a basis for the technique of qualitative reasoning (Bhaskar and Nigam, 1990; Sycara and Navichandra, 1989) on design objects have originated from the field of AI and have once more underlined the usefulness of this symbolic technique.

In this work it is shown that dimensional analysis can be used to solve the evaluation problem which frequently occurs during design synthesis when choosing among various design alternatives. Since dimensional analysis relies on the functional modeling of the design object, the applicability of the method is restricted to the concept and limitations of functional modeling in the CAD process (Kuttig, 1993). The implementation of this symbolic method could represent an useful enlargement to existing CAD systems by creating more abstract, symbolic information created from the physical descriptions of the design object contained in the database of the solid modeler. It is important to note that the existing prototype could simply be incorporated into the application programming interface of commercially available CAD/CAE systems (SDRC, 1993) without imposing any changes of the currently valid CAD/CAE paradigms or technologies (Hoschek, 1993; Hoschek and Dankwort, 1994).

To introduce the theoretical concept of the evaluation hypothesis based on dimensional analysis heavily used in the later sections, the evaluation problem of technical design objects as one of the key problems of design analysis and its place in the design process during design synthesis is briefly introduced in section 1. Section 2 states the key ideas of the evaluation hypothesis. Section 3 contains the necessary proofs and section 4 presents the most important properties which can be derived from these proofs. Section 5 gives two short analytic engineering examples to demonstrate the suggested technique. Section 6 presents the current status of the implementation of the suggested symbolic front-end and closes with an outlook on further conceptual developments.

In the following, the general framework of the design process is described and the evaluation problem inherent to the design process is identified.

1.1. DESIGN PROCESS

In a simplified view the design process can be understood as a sequence of more or less related decisions. These decisions affect the selection of a design topology as well as the selection of appropriate sizes of the related design parameters x_1 to x_n which describe the selected topology with sufficient precision.

It is evident that at almost every moment during the design process, design alternatives have to be evaluated with respect to the design evaluation criteria. How these decisions based on evaluation techniques are made is therefore of crucial importance for the sequence of events in the design process. For this reason, the problems underlying the construction of evaluation models will be investigated further in the following section.

1.2. EVALUATION PROBLEM

If one accepts the principle of decomposition of a general goal into several smaller subgoals and the aggregation of evaluations components into a global evaluation respectively, then an evaluation model based on such an assumption can only be a valuable tool for decision making in the design process if at least acceptable answers can be found for the following central questions:

- how to structure the used goal criteria hierarchy?
- how to determine the evaluation of various distinct goal criteria?
- how to aggregate multiple goal criteria into one single goal criterion?

Due to the lack of a formal methodology providing answers to these fundamental questions, most classical decision making models require these questions to be answered by a human decision maker (Hwang and Yoon, 1981). The task of the human decision maker is to establish the *description* graph of the design object, to determine the *evaluation* graph and then find the corresponding *mapping* of the description onto the evaluation. This is shown in Figure 1. The influence of the

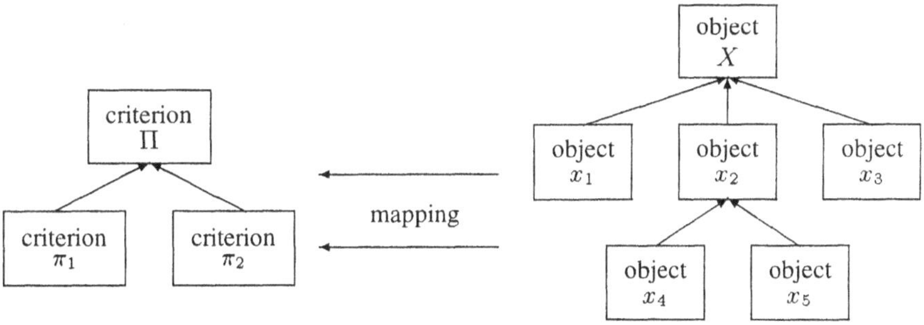

Figure 1. Object description space X, mappings and evaluation space Π.

decision maker leads to the central question of to what extent a decision reflects the personal beliefs of the decision maker or whether an evaluation should be the unique property of the design object. This issue will be discussed in the following section.

2. Motivation

From a comparative analysis of some existing decision theories and evaluation models (Rudolph, 1995) it can be concluded that:

- Under the assumption that an objective evaluation exists in general, it may not depend on the arbitrarily chosen definitions of physical units and therefore has to be *dimensionless*.
- A reproducible and objective evaluation can only exist if it is based on and derived from some type of law which has to be *dimensionally homogeneous*.
- An evaluation method should turn into exact physics and be consistent in the case of *complete* knowledge about a design object.

A universal method to construct dimensionless quantities from dimensionally homogeneous equations is given through the Pi-Theorem, which will be presented in the next section. To ease the understanding of the introduced model, from now on the following terminology will be used as shown in Figure 2, which is essentially the same diagram as shown in Figure 1. In Figure 2 the x_i and X represent the description, while π_j and Π represent the corresponding evaluation. The mapping is represented by φ_0 and φ_j, while φ_1 and φ_3 are the aggregation functions of the partial evaluations and the partial descriptions respectively.

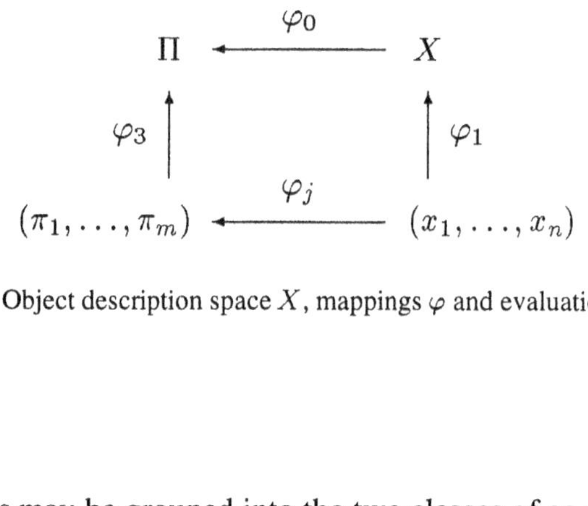

Figure 2. Object description space X, mappings φ and evaluation space Π.

3. Foundation

Physical quantities may be grouped into the two classes of so called primary and secondary quantities (Bridgman, 1922). Primary quantities are hereby quantities whose reference measurement is one of the base units of the employed unit system. Secondary quantities are derived from primary quantities by some function f. In this respect time (measured in s) and length (in m) are primary quantities, while velocity (in m/s) is a secondary quantity. Ratios of both primary and secondary quantities are invariant under scale transforms of the type $x' = \alpha x$ of physical units, e.g. $1[inch] = 0.0254[m]$. For the special properties of the function f the so called Product-Theorem can be proven (Bhaskar and Nigam, 1990; Bridgman, 1922):

Product-Theorem. *Due to the invariance of ratios of physical quantities under scale transforms of physical units, it can be shown that the function f relating a secondary quantity to some appropriate primary quantities $x_1, \ldots, x_n$ is of the* *form*

$$f = C x_1^{\alpha_1} x_2^{\alpha_2} \cdots x_n^{\alpha_n} \tag{1}$$

with $n \in \mathbb{N}$, C *and the* $\alpha_{ji} \in \mathbb{R}$.

This property of secondary quantities is then used to prove the so called Buckingham- or Pi-Theorem (Bhaskar and Nigam, 1990; Bridgman, 1922):

Pi-Theorem. *From the existence of a dimensionally homogeneous and complete equation* f *of* n *physical quantities* x_i *the existence of an equation* F *of only* m *dimensionless quantities* π_j *can be shown*

$$\begin{aligned} f(x_1, \ldots, x_n) &= 0 \\ F(\pi_1, \ldots, \pi_m) &= 0 \end{aligned} \tag{2}$$

where $r = n - m$ *is the rank of the dimensional matrix constructed by the* x_i *and with dimensionless quantities* π_j *of the form*

$$\pi_j = x_j \prod_{i=1}^{r} x_i^{-\alpha_{ji}} \tag{3}$$

with $j = 1, \ldots, m \in \mathbb{N}$ *and the* $\alpha_{ji} \in \mathbb{R}$ *as constants. (Examples of dimensional matrices are given in the application section.)*

An evaluation represents a qualitative and quantitative measure of an object or process. As long as its representation form is still redundant, this redundancy can be eliminated without loss of information. Therefore an evaluation possesses the property of a redundancy-free representation form and is minimal in this respect (Rudolph, 1995). This means that the number of independent parameters can't be reduced any further. This is expressed in the hypothesis that *"any minimal description in the sense of the Pi-Theorem is an evaluation"*.

4. Evaluation

Using the proof of the Pi-Theorem, the following list of selected properties can be shown for the evaluation method (Rudolph, 1995):

- *Evaluation.* The problem of evaluation can be principally reduced to a problem of description. The problem of evaluation is solved exactly in those cases where a complete description exists.
- *Minimality.* The dimensionless product exponents form a basis in the sense of a linear vector space. The properties of a vector space basis like minimality is therefore also valid for fundamental systems of dimensionless products.

- *Granularity.* The addition of one more x_{n+1} to the original description set of $x_1, \ldots, x_n$ adds one more π_{m+1} and leaves the original set of evaluation components $\pi_1, \ldots, \pi_m$ unchanged. This property supports the experience of hierarchical refinement in the sequence of the design process.
- *Hierarchy.* Solving F for a specific π_j creates immediately a consistent hierarchy as shown in Figure 2. This property can be extended to multiple hierarchies.
- *Sensitivity.* The differential formulation of the model laws with $\pi_j = const$ is $d\pi_j = 0$. Differentiating equation (3) leads to

$$d\,\pi_j \;=\; \frac{\partial \pi_j}{\partial x_j} d\,x_j + \sum_{i=1}^{r} \frac{\partial \pi_j}{\partial x_i} d\,x_i \quad j = 1, \ldots, m \tag{4}$$

  and setting $d\,\pi_j = 0$ leads to the general form of an *iso-line* of an evaluation component. If only infinitesimal changes of two design parameters x_j and x_i are permitted, with all other changes equal to zero one obtains

$$\frac{\partial x_j}{\partial x_i} \;=\; \alpha_{ji} \frac{x_j}{x_i} \qquad \begin{array}{l} i = 1, \ldots, r \\ j = 1, \ldots, m \end{array} \tag{5}$$

  which is analogous to the expression derived in Bhaskar and Nigam (1990) for the purpose of *"qualitative reasoning"*.
- *Modularity.* The structure of the dimensionless products can be encoded into a topological matrix a. With additional use of the unit matrix I the coupling matrix $\kappa = a^T a$ is

$$\kappa \;=\; a^T I\, a \tag{6}$$

  This is an analogy to the construction of a stiffness matrix of a structure created from stiffness matrices of multiple finite elements, see also Table 4.

A few ideas on how the method can be tested is given by the fulfillment of the following selected statements and arguments (Rudolph, 1995):

- *Causality.* The evaluation Π is determined by the complete description X in the mathematical sense as a necessary and sufficient condition.
- *Invariance.* The evaluation Π is invariant under scale transforms of the physical units employed in the description X of the object or process.
- *Abstraction.* Since the evaluation Π is the property of a whole class of similar but well distinct objects in X, the mapping from X to Π is mathematically surjective and not injective.
- *Consistency.* Since the evaluation is generated by a mapping, the consistency over multiple hierarchy levels is guaranteed if the theory underlying the description hierarchy is consistent.

5. Application

The evaluation method is shown using two examples. The first example focuses on the demonstration of some of the properties of the method like the derivation of the evaluation components and the demonstration of hierarchy and consistency. The second example is taken from a publication in the area of AI (Bhaskar and Nigam, 1990), so that the link established by the mathematical formulation between engineering design evaluation and AI reasoning techniques becomes evident.

5.1. AUTOMOBILE

If one tries to evaluate the aerodynamic properties of an automobile like the one sketched out in Figure 3, the relevant physical quantities for the description of the

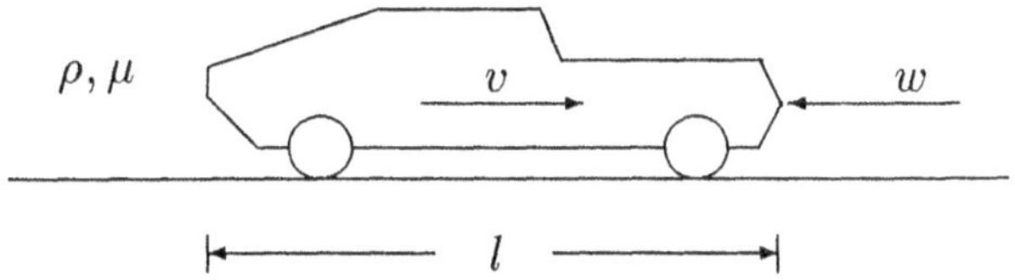

Figure 3. Relevant design parameters of car drag model.

underlying physical process construct the dimensional matrix shown in Table 1. From this dimensional matrix with dimension $n = 5$, rank $r = 3$ the following

TABLE 1. Dimensional matrix of car drag parameters.

Symbol	$[M]$	$[L]$	$[T]$	SI-Units	Meaning
l		1		m	characteristic length
v		1	-1	m/s	velocity of car
ρ	1	-3		kg/m^3	density of air
μ	1	-1	-1	$kg/m\ s$	viscosity of air
w	1	1	-2	$kg\ m/s^2$	drag of car

$m = n - r = 2$ dimensionless products π_1 and π_2 can be derived. According to the

$$\pi_1 = \frac{w}{\rho v^2 l^2} \qquad (\equiv c_w) \tag{7}$$

$$\pi_2 = \frac{v l \rho}{\mu} \qquad (\equiv Re) \tag{8}$$

complete description $f(x_1, \ldots, x_5) = 0$, a relation of the form $F(\pi_1, \pi_2) = 0$ of two dimensionless variables exists. With reference to the general scheme in Figure 2, this interrelation is shown in Figure 4. While φ_1 is given by the chosen

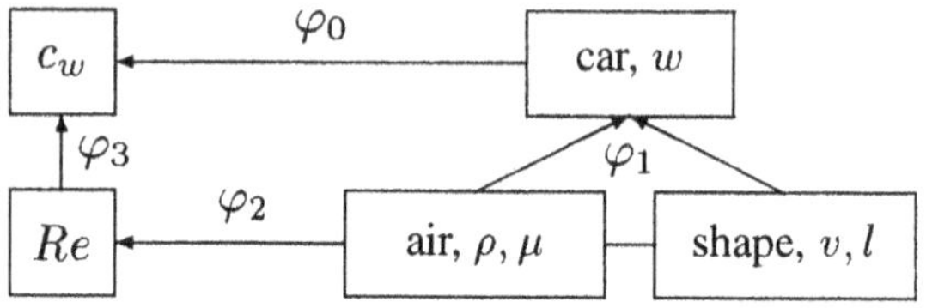

Figure 4. Description and evaluation graph of car drag w.

explicit form $w = f(v, l, \rho, \mu)$ of the implicit formulation of f, φ_0 and φ_2 are determined by equation (7) and (8). Thus only φ_3 still needs to be determined. This is mostly done by function approximation of experimental or numerical data for the whole class of geometrically similar objects, as shown in Figure 5 for the class of

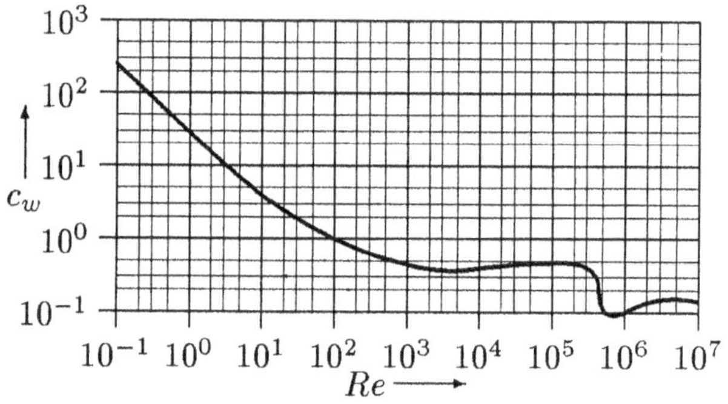

Figure 5. Drag coefficient c_w of spheres.

spheres (Zierep, 1972). The approximation and visualization of experimental or numerical data using φ_3 instead of φ_1 is advantageous, since the number of independent parameters is reduced and the interpretation of the obtained relationship more general (Kline, 1986).

5.2. PRESSURE VALVE

The modeling of the pressure valve is done according to the presentation in Bhaskar and Nigam (1990). There the whole mechanism as shown in Figure 6 is modeled in two parts: (A) the pipe, using a functional description of $f_A(\rho, a, p_i, p_o, q) = 0$, (B) the orifice, using a functional description of $f_B(k, x, p) = 0$. The corresponding dimensional matrices are named A and B, see table 2. The interaction of both

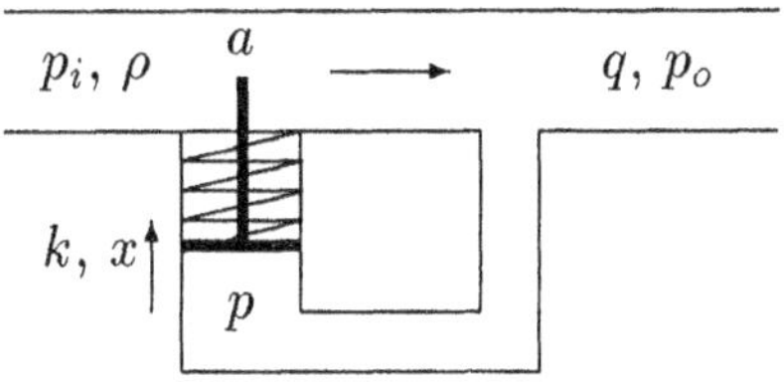

Figure 6. Relevant design parameters of pressure valve.

TABLE 2. Dimensional matrices of pressure valve components A nd B.

A	[M]	[L]	[T]	SI-units	Meaning
ρ	1	-3		kg/m^3	density
a		2		m^2	area
p_i	1	-1	-2	N/m^2	pressure in
p_o	1	-1	-2	N/m^2	pressure out
q		3	-1	m^3/s	stream flow

B	[M]	[L]	[T]	SI-units	Meaning
k	1		-2	kg/s^2	spring constant
x		1		m	elongation
p	1	-1	-2	N/m^2	local pressure

components is modeled using (C) two coupling conditions, with functional descriptions $f_{C1}(p, p_o) = 0$ and $f_{C2}(x, a) = 0$. The corresponding dimensional Matrices are named C1 and C2, see table 3. Using these four matrices, the following dimensional products can be constructed, which in Bhaskar and Nigam (1990) are also called *ensembles*

$$\pi_{1A} = \frac{q\,\rho^{1/2}}{a\,p_i^{1/2}} \tag{9}$$

$$\pi_{2A} = \frac{p_o}{p_i} \tag{10}$$

$$\pi_{1B} = \frac{x\,p}{k} \tag{11}$$

$$\pi_{1C} = \frac{p}{p_o} \tag{12}$$

$$\pi_{2C} = \frac{x}{a^{1/2}} \tag{13}$$

TABLE 3. Dimensional matrices of functional coupling conditions C1 and C2.

$C1$	[M]	[L]	[T]	SI-units	Meaning
p	1	-1	-2	N/m^2	local pressure
p_o	1	-1	-2	N/m^2	pressure out

$C2$	[M]	[L]	[T]	SI-units	Meaning
x		1		m	elongation
a		2		m^2	area

Since the five ensembles have some of the variables in common, they are not independent from each other and the coupling relationship can be drawn as an undirected graph as shown in Figure 7.

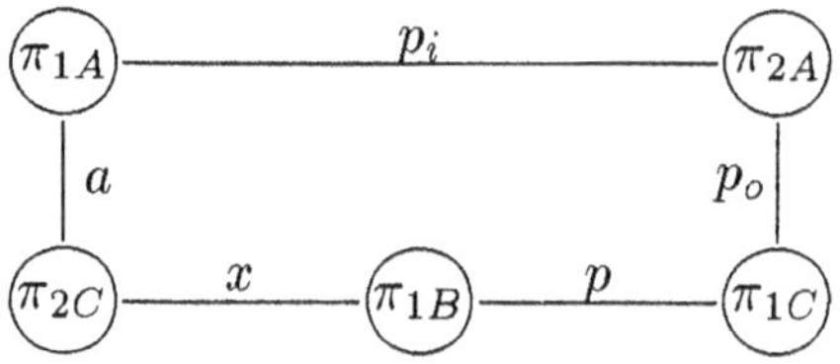

Figure 7. Undirected graph of ensembles (Bhaskar and Nigam, 1990).

With the calculus derived in equation (5), the qualitative behavior of p_o due to small changes in p_i can now be determined with a strategy called *"qualitative reasoning"*. Using the chain rule, the sign of the derivative $\frac{\partial p_o}{\partial p_i}$ can be determined, when expanding the derivative along the path $\pi_{1C} \rightarrow \pi_{1B} \rightarrow \pi_{2C} \rightarrow \pi_{1A}$ in the undirected graph.

$$\frac{\partial p_o}{\partial p_i} = \frac{\partial p_o}{\partial p}\frac{\partial p}{\partial k}\frac{\partial k}{\partial x}\frac{\partial x}{\partial a}\frac{\partial a}{\partial q}\frac{\partial q}{\partial p_i} < 0 \tag{14}$$

The coupling matrix κ of the design variables constructed using equation (6) depends only on the product form of the dimensionless products and is shown in Table 4.

In analogy to the construction of a stiffness matrix in cartesian coordinates (Argyris and Mlejnek, 1986) of a structure created from stiffness matrices of multiple finite elements in natural coordinates, the five dimensionless variables can be

TABLE 4. Design coupling matrix κ.

						q	ρ	a	p_i	p_o	p	x	k	
						1	1	1	1					π_{1A}
									1	1				π_{2A}
a											1	1	1	π_{1B}
										1	1			π_{1C}
								1				1		π_{2C}
	1					1	1	1	1					
		1							1	1				
I			1								1	1	1	Ia
				1						1	1			
					1			1				1		
	1					1	1	1	1					
	1					1	1	1	1					
	1				1	1	1	2	1			1		
	1	1				1	1	1	2	1				
a^T		1		1					1	2	1			κ
			1	1						1	2	1	1	
			1		1			1			1	2	1	
			1								1	1	1	

interpreted as the natural degrees of freedom of the design object. In this representation the off-diagonal elements are indicators for the implicit coupling of the design variables.

In the following, a short description of the current implementation of the suggested symbolic CAD-front-end based on the suggested techniques will be given.

6. Symbolic CAD Front-End

Using the previous example of the pressure valve design, the currently implemented features of the suggested symbolic front-end for CAD/CAE systems are shown in Figure 8. Since the UNIX-based, network-transparent software programmed in Motif/C is not yet fully incorporated into the CAD application programming interface, a rough sketch of the pressure valve and its physical design parameters involved in the modeling are displayed as a postscript image in the upper right corner. The engineer can then select the appropriate variable descriptions in a database provided by the system, as shown in the upper left corner (see window *Database Form*).

By selecting the appropriate design variables, the dimensional matrices of the two components of the pressure valve and their topological coupling conditions are constructed (see the three windows *Analyze Form*), and, according to equation 3, the associated dimensionless groups are automatically generated. The symbolic in-

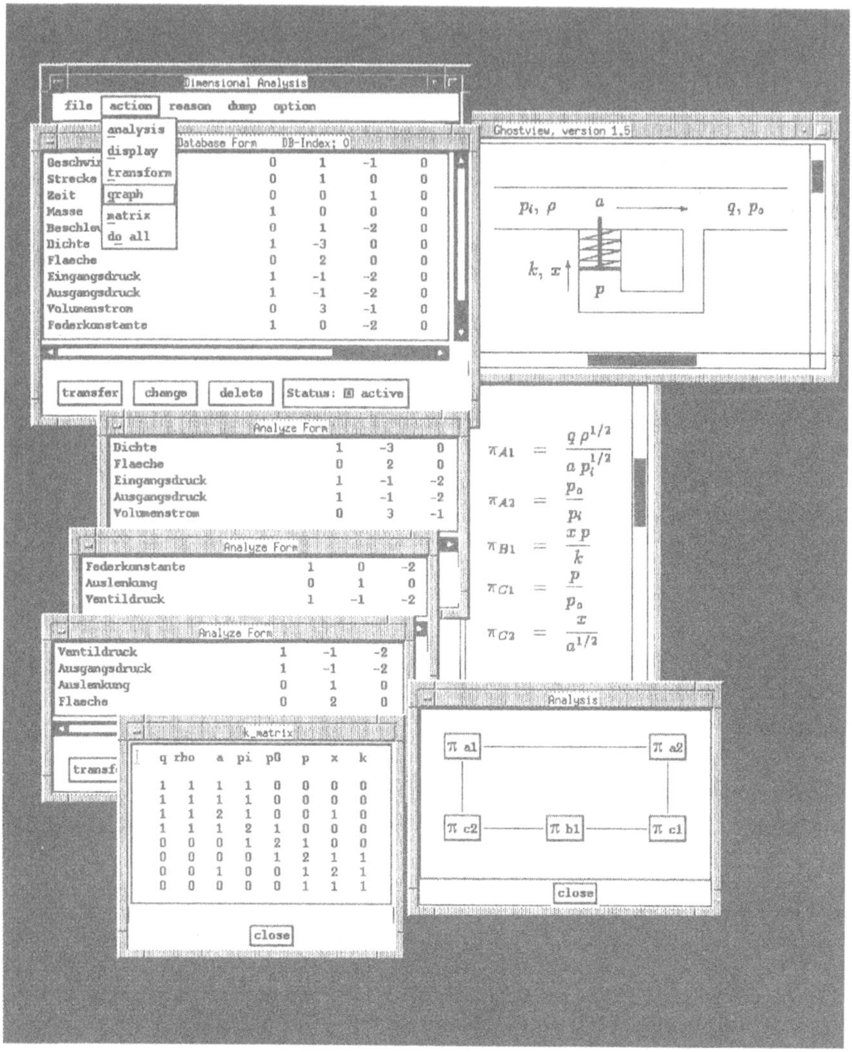

Figure 8. Symbolic CAD front-end prototype.

formation of the building structure of the dimensionless groups is then converted into a postscript format and displayed on the screen as well (see background window in the middle). Using this symbolic information, the coupling matrix according to equation 6 (see window *k_matrix*) and the corresponding undirected graph (see window *Analysis*) are also generated by the system.

The manual selection of the design variables and the generation of the symbolic information takes the designing engineer about less than a minute. Once incorporated into the application programming interface, this step can be further automated. This means that it could be imagined that the current design parameter values $x_1, \ldots, x_n$ are extracted and continuously updated from the database of the solid modeler of the CAD system. The information automatically created by the system in this way would provide the engineer with valuable qualitative insights into the interdependencies of his design. In this respect, the closed loop in the undirected graph window indicates the existence of possible feedback (Bhaskar and Nigam, 1990).

Despite the relative little time of the availability of this software tool and the little experience gained with it until now, the potential of the approach of combining symbolic properties of *finite components* in analogy to the combination of numerical properties in classical *finite element* methods seems apparent. Due to the fact that the theoretical basis of the approach relies on the traditional engineering method of dimensional analysis, any other techniques relying on this notation, such as qualitative reasoning techniques (Bhaskar and Nigam, 1990) or engineering design using similar size ranges (Pahl and Beitz, 1993), both based on dimensional analysis, might be straightforwardly added. This concept seems therefore to be worthwhile further investigation.

7. Conclusion

By interpretation of the postulated evaluation hypothesis, it is most interesting to note that it can be shown that in a rigorous sense an evaluation is only possible if a complete description in form of functional relationships of the design parameters are known. Since complete functional descriptions are generally difficult to obtain, incomplete descriptions in the case of complex design problems will not allow to evaluate missing technical aspects nor give a complete set of dimensionless products. Therefore not all coupling between design variables will be detected and originally coupled design spaces will be looked at as orthogonal. This fact however is not a specific weakness of the suggested approach, but reflects the fact of incomplete modeling of physical phenomena only. The coupling matrix κ in Table 4 will reflect these facts by less off-diagonal entries.

Even though the method put that much emphasis on a clear understanding and a complete description of the problem, dimensional analysis based on the Pi-Theorem has always been a valuable tool for engineers facing and investigating complex technical problems. Proving the completeness of the relevance list $x_1, \ldots, x_n$ of parameters is restricted to areas of "sharp" physical knowledge but the method with its property as a mathematically necessary condition can't lead to formal contradictions by itself. This imposes no principal restrictions on the use of the method in areas of "unsharp" physical knowledge.

Of further interest is also the question, what consequences will be caused by the necessary selection of one specific set of $\pi_1, \ldots, \pi_m$ out of the mathematically equivalent $(n-r)$-parametric solution space of dimensional matrices. Due to the mathematical formulation, the link between classical similarity methods, design evaluation methods and qualitative reasoning techniques could be shown. Such a consistent formulation might ease algorithmic and implementation aspects of future design systems combining these features.

A future design system could possibly exploit detection of similarity for the creation of associations or analogy conclusions (Kodratoff, 1990). Further, the identification of weak coupling of design variables could be used to develop decomposition strategies for automatic design optimization. The investigation of such possibilities might be a first step to make one day more efficient or even more intelligent design support systems a reality.

8. Acknowledgments

The author wants to thank Peter Hertkorn for his programming support of the Motif/C based interactive software interface and thanks Robert Schütz for the partial implementation of the underlying algorithms. The financial support of this work by the *Deutsche Forschungsgemeinschaft (DFG)* and the German interdisciplinary research group *Forschergruppe im Bauwesen (FOGIB) "Ingenieurbauten — Wege zu einer ganzheitlichen Betrachtung"*, is greatly acknowledged.

References

Andreasen, M.: 1992, The theory of domains, *Technical Report CUED/G-EDG/TR11*, Engineering Department, Cambridge University, Cambridge.
Argyris, J. and Mlejnek, H.-P.: 1986, *Die Methode der Finiten Elemente I*, Vieweg Verlag, Braunschweig.
Bhaskar, R. and Nigam, A.: 1990, Qualitative physics using dimensional analysis, *Artificial Intelligence*, **45**, 73–111.
Bridgman, P.: 1922, *Dimensional Analysis*, Yale University Press, New Haven.
Dolinskii, I.: 1990, Use of dimensional analysis in the construction of mechanical assemblies for optical instruments, *Soviet Journal of Optical Technology*, **57**(8), 512–514.
Dym, C.: 1994, *Engineering Design. A Synthesis of Views*, Cambridge University Press, Cambridge.
Hoschek, J. (ed): 1993, *Was CAD-Systeme wirklich können*, Teubner, Stuttgart.
Hoschek, J. and Dankwort, W. (ed): 1994, *Parametric and Variational Design*, Teubner, Stuttgart.
Hwang, C.-L. and Yoon, K.: 1981, *Multiple Attribute Decision Making*, Lecture Series in Economics and Mathematical Systems, Springer-Verlag, Berlin.
Kline, S.: 1986, *Similitude and Approximation Theory*, Springer-Verlag, New York.
Kloberdanz, H.: 1991, *Rechnerunterstützte Baureihenentwicklung*, Fortschrittsberichte Reihe 20, Nummer 40, VDI-Verlag, Düsseldorf.
Kodratoff, Y.: 1990, Combining similarity and causality in creative analogy, *Proceedings ECAI-90*, pp. 398–403.

Kuttig, D.: 1993, Potential and limits of functional modelling in the CAD process, *Research in Engineering Design*, **5**, 40–48.
Pahl, G. and Beitz, W.: 1993, *Konstruktionslehre*, Springer-Verlag, Berlin.
Rudolph, S.: 1995, *Eine Methodik zur systematischen Bewertung von Konstruktionen*, VDI Fortschrittsberichte, Reihe 1, Nummer 251, VDI-Verlag, Düsseldorf (In German. An English version entitled: *A methodology for the systematic evaluation of engineering design objects* is available on request by email to rudolph@isd.uni-stuttgart.de.
SDRC Master Series: 1993, *User's Guide: Model Solution and Optimization*, Structural Dynamics Research Corporation, Milford, Ohio.
Suh, N.: 1990, *The Principles of Design*, Oxford Press, New York.
Sycara, K. and Navinchandra, D.: 1989, Integrating case-based reasoning and qualitative reasoning in engineering design, *Proceedings Applications of AI in Engineering*, pp. 231–250.
ten Hagen, P. and Tomiyama, T.: 1991, *Intelligent CAD Systems I*, Springer-Verlag, New York.
VDI Guideline-2221: 1987, *Systematic Approach to the Design of Systems and Products* (trans. of the German edn 11/1986), VDI-Verlag, Düsseldorf.
Zierap, J.: 1972, *Ähnlichkeitsgesetze und Modellregeln in der Strömungslehre*, Braun Verlag, Karlsruhe.

10

PERFORMANCE EVALUATION METHODS IN DESIGN: DISCUSSION

HANS GRABOWSKI
University of Karlsruhe, Germany

Product development has a close connection with the theory of decision making. That theory deals with the human behaviour of decision making in its social context. It is an interdisciplinary branch of economics. The interdisciplinarity is caused by the influences of mathematics, psychology and sociology.

Von Neumann and Morgenstern published their classic work on game theory called *Theory of Games and Economic Behaviour* in 1944. In their work, the authors prove the similarity between the behaviour of an economics system and the mathematical representation of certain technical games. While classical decision theory deals with the behaviour of individuals in deciding, game theory developed as an independent discipline which studies the conflict of interest between at least two decision makers.

In contrast to game theory, stands the area of vector optimization, which is also known under the terms multiobjective optimization or multicriteria optimization. It deals with the problem of multiple goals of one single individual with the help of multicriteria decision models. Multicriteria optimization also originated from economics and has been used by engineering disciplines, especially by civil engineering.

Decision theory distinguishes between normative, descriptive and prescriptive approach. Normative decision theory assumes the concept of an optimal decision. Its behaviour is described with a formal logic of decision making. Descriptive decision theory assumes, in contrast to the closed model of normative decision theory, an open model of decision making. This open model takes the creation of the decision criteria and the insufficient capacity of human information processing into account. This model explains why in complex decision making situations, the individual does not always decide in an optimal way. Prescriptive decision theory does not aim to develop its own models of decision making but supports the decision maker through the conservation of consistency.

There is a growing unification of decision theory and the theory of cognitive problem solving processes is developing. While decision theory focuses on the problem of the solution process as preparation for the decision, cognitive theory focuses on the decision as the end of the problem solving process. It is implicitly assumed that the problem definition also contains a request for optimization so that a requirement for a decision is explicit. The almost identity of the decision and problem solving processes is clarified by a schema of the phases of the decision process which distinguishes six steps:

- Identification of the problem,
- Acquisition of necessary information,
- Development of possible solutions,
- Evaluation of the solution,
- Selection of a strategy for the action,
- Action followed by learning and revision.

This phase schema of the decision process is very similar to the model of the systematic procedure for the problem solving process which was integrated into the phase oriented design process in design methodology. The iterative character of the problem solving process is explicitly shown in the last step. As a consequence, the phase schema develops a cyclic process which is called "cybernetic control cycle" or "cognitive program" by psychologists. Figure 1 shows this cybernetic control cycle in a so-called "Test-Operate-Test-Exit" (TOTE) unit.

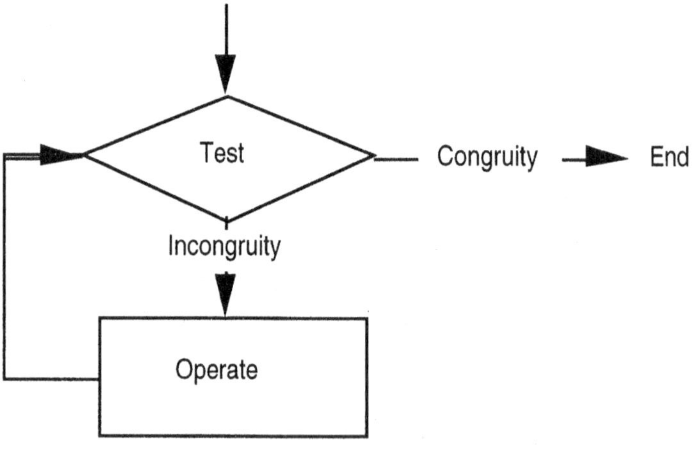

Figure 1. TOTE unit.

The program of a decision and optimization process is caused by an incongruity between the requirements which are valid for a decision and the current situation. With respect to the design process, the situation is described by the description of the design problem or by a first solution of this problem. In both cases an analysis (test) shows the necessity to correct detected errors. This cycle of analysis and synthesis is repeated until a sufficient (congruous) solution was found (exit).

With the help of low-level and high-level TOTE units, which represent strategic or tactical units respectively of a problem solving process plan, a hierarchical structure of TOTE units, can be designed where specific TOTE units can contain other TOTE units. In addition to the iterative character of the decision making and problem solving process, a recursive element occurs. This recursive element realizes a higher level goal by satisfying several low-level sub-goals. This produces problem structuring.

The efficiency of the product development process would be improved if realistic goals of the design intent could be derived from the requirement specification at an early stage of design. In most cases of product planning some idealistic concepts about the future product very often cannot be satisfied. The designer now has to answer the question which of these goals are important and must be reached by the design and which of them are not that important and can be neglected. A clarification of the task in advance would augment the efficiency of the product development process.

A number of approaches have been developed as part of design methodology to deal with these multiple and often conflicting goals. These approaches fall into two clearly distinguishable groups. The first group consists of methods which use weights attached to goals to indicate their relative importance. This approach has obvious difficulties associated with how these weights are arrived at and how they should change with a change in the available information. Often weights are implied but never explicitly stated. The second group consists of methods which explicitly do not use weights—the Pareto methods. These methods delay the final decisions to a later phase.

In many cases the weights only represent the importance of the respective design goals whereas a relation to the actual instances of the weighted parameters is not given. Design evaluation methods must consider both facts, the relative importance of a specific goal and the possible range of its values.

So, in the decision making process *evaluation methods* are necessary between every design stage because if a design solution cannot be evaluated no progress in the design process is possible. Three papers dealing with evaluation methods from their respective points of view were presented and discussed in this session. The paradigm form follows function or function follows form and the fomalization of that paradigm is no longer the point at issue as it was in the past. Most of the authors introduced terms like *context* explicitly to express that the concept of causality between form and function does not lead to practical and "good" design solutions. So each author had to incorporate and to formalize more semantics in the data models which have to be evaluated. The requirements specification has to be formalized so that formal methods can evaluate a design solution against the requirements.

Requirements can be specified quantitatively or qualitatively. Every evalution method presented supports quantitative and qualitative evalution explicitly. Differences could be seen in the support of the design process itself and the design stages in which the evalution methods are embedded. Kalay et al. and Rodgers et al. embedded their evaluation concept in a design process while Rudolf's evaluation method supports evaluation at a unique design stage.

The paper *A Performance Based Paradigm of Design* by Yehuda Kalay and Gianfranco Carrara proposes an approach which is called performance based design. The main idea is that the relationship between form and function is not causal but contextual which means that finding a design solution does not only depend on the relationship between form and function but on additional information such as social, physical or cultural context information. We can say that contextual information is all the information which is not form and function. The computational model explicitly represents function and form, as well as the context. The main elements of the model are a structured set of goals, representing the functional requirements, a set of solutions, a context and a set of evaluators.
In the paper *On a Symbolic CAD Front End for Design Evaluation Based on the Pi-Theorem* by Stephan Rudolph, an evaluation method which is based on the Pi-theorem of physics is presented. The basic idea to the evaluation problem is to transform a given functional relationship of design parameters into a dimensionless evaluation form. The design parameters are described in an object description space x, which is a homogeneous and complete description f of n physical quantities x_i: $f(x_i, \dots, x_n)$. The dimensionless representation is described by an equation F of m dimensionless quantities π_j. The mapping is performed by a mapping function φ. Using the theory of

Pi-theorems it can be proved, that the dimensionless graph of $F(\pi_1, \dots, \pi_m)$ is complete and consistent. It can be shown, that this evaluation method has theoretical properties such as minimality, consistency, granularity, hierarchy, sensitivity and modularity and provides a sensitivity analysis of the design parameters. In addition to that the Pi-theorem shows, that the evaluation problem can be reduced to the object description space x,: $f(x_i, \dots, x_n)$. Rudolph demonstrated the practical feasibility of this method with a prototype of a symbolic CAD front end that has been integrated with a CAD system. The practical benefit is based on supporting the designer by automatic generation and display of the dimensionless groups. In order to obtain this, the designer only has to select the appropriate design variables and to construct the topological coupling conditions. For the graphical output the symbolic information of the dimensionless groups is converted into a graph of the coupling of the design variables on the screen.

In the paper *A Formal Method for Assessing Product Performance at the conceptual Stage of the Design Process* by Paul Rodgers et al., the interrelation between form of an intended product and the context of its use is the driving factor of design, where context represents product behaviour from users' point of view. In the paper, context is understood to determine the design problem.

There are two different classes of design processes, the unselfconscious and selfconscious process. The first is classified by the designer operating dircctly on an actual form, determining misfits and correcting them.

One characteristic of the latter is that the designer is not able to directly observe the ensemble. The designer investigates, explores and researches the actual context and constructs a mental picture of it. With this mental picture of the context and with a representation of the product, the designer makes a mental picture of the ensemble in order to try to determine potential misfits.

In the approach to product performance assessment presented, attribute prediction plays an important role as well as the extraction (recognition) of product characteristics. Characteristics are inherent properties of any product, independent of the product use. They can be determined purely from their representation.

The methodology was implemented in an expert system where the following procedure for product performance assessment is used:

1. Define the attributes in users' terms,
2. Determine the model of each attribute (design of the related rule base),
3. Define the combination function.

The implemented system evaluates the proposed design by evaluating the attribute model with the help of the defined combination function. It answers the main criterion and returns an objective value determined for it.

All presentations explained that it is a unique problem to extract the criteria against which a design is evaluated. The discussion on this topic led to the conclusion that the product requirements represent these criteria and that this is the reason why the product requirements, which must contain information of all stages of the product life cycle and especially different views on the product (design view, users' view), have to be formalized in a computable model. However there still remains the problem of weighting different requirements if weights are used. An absolute weighting of the design parameters (requirements) is not useful and it does only makes sense to weight the requirements relative to each other. The question how to represent "trade-offs" leads to the result that trade-offs too can and should be modeled by requirements and that an evaluation system should not only be used in order to justify a design but also to aid the designer in the decision making process.

In which way requirements can change throughout the design process was another important topic of the discussion and was discussed vigorously. For example, a designer is given the task of designing a bridge over a river and because of certain problems and restrictions, the design ends up as a tunnel. There was consensus that new requirements can appear during the design process but the point at issue was the nature of requirements, the abstraction level of the represented information and, of course, the than open problem of formalizing requirements. The inadequacy in this example was that the problem definition was not abstracted enough, which means that during the definition of the design task the product (bridge/tunnel) has been specified and not the problem represented by the requirements. The designer has to start at a fixed point. That point is the main function representing the intended design. If the main function of the design, to transport objects from one side of a river to the other, is given then the intended design will not be preordained and it is possible that requirements, especially costs etc., can change over time without necessarily producing a totally different product.

PART FOUR

Formal Support Methods in Design

11

FORMAL CONCEPT ANALYSIS IN DESIGN

MIHALY LENART
University of Kassel, Germany

Abstract. Formal concept analysis is a research method using set theoretical models of concepts and their hierarchical orderings. The model is based on concept or Galois lattices whose application for analyzing design contexts has been proposed earlier by Ho (1982a). Although concept lattices have been widely used for analyzing contexts in various areas, such as music, social sciences, or cognitive science, the analysis of design contexts by concept lattices has not gained acceptance. In particular, it has not been recognized yet that Hasse diagrams, a representation of concept lattices by line diagrams, can not only help to visualize important hidden properties of a design context, but also provide a new kind of analytical tool that can support the decision making process. The paper shows that Hasse diagrams contain all information of the underlying context and reveal inherent structural dependencies not captured by any other graphical representation of the given context. It shows also how concept lattices, or rather their Hasse diagrams, can be used for supporting the design process in general and for analyzing design contexts in particular.

1. Introduction

Many attempts have been made to describe design contexts exactly by formal, mathematical means. The purpose of such description is to understand relations or dependencies between design objects and to develop computer programs for automating certain parts of the design process. One way to characterize a design context is to describe design objects as elements of a set and relations and/or operations on these elements. The result is an algebraic model of the design context. There are numerous such models and the formal description of a context we are using here is probably the simplest one. Nevertheless, this simple model can capture basic features of a design context. The same model has been used also for Q–analysis that was developed by Atkin (1974) in the early 1970s. It was quite popular and widely used for the analysis of design contexts in the late 1970s and early 1980s. It turned out, however, that some of the information get lost in the process and other algebraic tools, such as concept lattices, are superior to or more adequate than Q–analysis for analyzing design concepts.

In the early 1980s, Ho has proposed the use of new algebraic tools, in particular set and category theoretical ones, for a formal description of the design process (Ho, 1982a; Ho, 1982b; Ho, 1982c). One of these tools was concept or — as Ho called them — Galois lattices. Although Ho's proposal was novel to design, concept lattices have been widely used previously in other social and scientific areas. The ground work for these applications has been laid down by Wille (1981; 1983; 1984; 1987; 1992). He has developed not only the lattice theoretical foundations but also various lattice generation and representation methods. Beyond the theoretical part, Wille and his co–workers have demonstrated the scope and the power of these methods by numerous applications. The aim of this paper is to utilize Wille's methods for design purposes.

2. Concept Lattices

Concept analysis is based on a common data type called (formal) context. From a context several mathematical objects, such as concept lattices, can be derived.

Definition 2.1 *A* context *is a triple* (O, P, γ) *where* O *and* P *are sets and* $\gamma \subseteq O \times P$ *is a binary relation between* O *and* P.

The elements of O and P describe entities of a given design situation or problem. According to March (1982), we need *at least* "two descriptive systems" in order to be able to describe design tasks or processes. In many cases two distinctive systems are also sufficient to describe relevant characteristics of a design context. A simple example is given in Figure 1 describing building structures and their main characteristics.

The two sets O and P can be chosen arbitrarily, however, for all practical purposes the "two distinctive description systems" mean that we usually have two sets of different kinds of entities, such as facts and values, objects and features, forms and functions, or locations and activities (Ho, 1982a). Keeping this in mind, we can assume — without loss of generality — that O is a set of objects and P is a set of properties. If an object $o \in O$ has a property $p \in P$, i.e. $(o, p) \in \gamma$ then we write $o\gamma p$. The context (O, P, γ) can be represented by a matrix with the objects heading the rows and the properties the columns. If an object o_i has the property p_j then we put an x into the field (i, j) and leave it empty otherwise (see Figure 1). For all subsets $X \subseteq O$ and $Y \subseteq P$ we define the following *derivation operations* represented by "prime":

$$X \longmapsto X' = \{p \in P \mid o\gamma p \; forall \; o \in X\} \tag{1}$$

$$Y \longmapsto Y' = \{o \in O \mid o\gamma p \; forall \; p \in Y\} \tag{2}$$

Let $\wp X$ denote the power set of X and $\wp Y$ the power set of Y. Then the pair (F, G) of mappings $F : X \wp X'$ and $G : Y \wp Y'$ are said to form a *Galois connection* between $\wp X$ and $\wp Y$ for which the following properties hold (Birkhoff, 1967):

$$X_1 \subseteq X_2 \; implies \; X_2' \subseteq X_1' \; for \; X_1, X_2 \subseteq O \tag{3}$$

building structure \ property	good fire proof	good sound proof	high constr. costs	short contr. time	low weight	long span	
cast in place	x	x				x	a
steel frame			x	x	x	x	b
wood frame				x	x		c
aluminum frame			x	x	x		d
masonry	x	x					e
prefab. concrete	x	x		x		x	f
	1	2	3	4	5	6	

Figure 1. A context describing relations between building structures and structural properties.

$$Y_1 \subseteq Y_2 \; implies \; Y_2' \subseteq Y_1' \; for \; Y_1, Y_2 \subseteq P \tag{4}$$

$$X \subseteq X'' \; and \; X \subseteq X''' \; for \; X \subseteq O \tag{5}$$

$$Y \subseteq Y'' \; and \; Y \subseteq Y''' \; for \; Y \subseteq P \tag{6}$$

$$(\bigcup_{t \in T} X_t)' = \bigcap_{t \in T} X_t' \; for \; X_t \subseteq O(t \in T) \tag{7}$$

$$(\bigcup_{t \in T} Y_t)' = \bigcap_{t \in T} Y_t' \; for \; Y_t \subseteq P(t \in T) \tag{8}$$

The relations above show a natural 'duality' between objects and properties.

Definition 2.2 *For a given context (O, P, γ), we call a pair (A, B) a* concept *if $A \subseteq O, B \subseteq P, A = B'$ and $B = A'$. A and B are called the* extent *and the* intent *of the concept (A, B) respectively.*

Figure 2 shows all the 11 concepts of the context of Figure 1.

Let us denote the set of all concepts of the context (O, P, γ) by $\mathbf{L}(O, P, \gamma)$. It can be shown that there is an ordering structure on $\mathbf{L}(O, P, \gamma)$ defined by the set inclusion relation: For any two concepts (A_1, B_1) and (A_2, B_2) we have $(A_1, B_1) \leq (A_2, B_2)$ if and only if $A_1 \subseteq A_2$ which is equivalent to $B_2 \subseteq B_1$ by 4. In this case we call (A_1, B_1) a *subconcept* of (A_2, B_2) and (A_2, B_2) a *superconcept* of (A_1, B_1). For the subconcept–superconcept relation we denote the ordered set $(\mathbf{L}(O, P, \gamma), \leq)$ by $\underline{\mathbf{L}}(O, P, \gamma)$.

	B'=A →	A'=B
1	0	1,2,3,4,5,6
2	b	3,4,5,6
3	f	1,2,4,6
4	a,f	1,2,6
5	b,d	3,4,5
6	b,f	4,6
7	a,b,f	6
8	a,e,f	1,2
9	b,c,d	4,5
10	b,c,d,f	4
11	a,b,c,d,e,f	0

Figure 2. The concepts derived from the building structure context.

In order to formulate the basic theorem showing that $\underline{\mathbf{L}}(O, P, \gamma)$ is a complete lattice, we need to introduce some lattice theoretical notations [1].

Definition 2.3 *A* partially ordered set *P (or for short* poset*) is a set with a binary relation* $\leq$ [2] *such that for all* $x, y, z \in P$ *we have* reflexivity: $x \leq x$, antisymmetry: *if* $x \leq y$ *and* $y \leq x$ *then* $x = y$, *and* transitivity: *if* $x \leq y$ *and* $y \leq z$ *then* $x \leq z$

Such a (reflexive, symmetric and transitive) relation is called a *partial order* on P. If any two elements of a poset are $\leq$–related we say that the poset is a *totally ordered* set or a chain.

Definition 2.4 *If* X *is a subset of a poset* P *(i.e.* $X \subseteq P$*), then* $a \in P$ *is an* upper bound *of* X *if for each element* $x \in X$ $a \geq x$. *The* least upperbound *or supremum of* X *denoted by* $supX$ $(\vee X)$ *is the smallest upperbound of* X. *(It can easily be shown that each subset has at most one.) Similarly, an element* $b \in P$ *is a* lower bound *of a subset* $X \subseteq P$ *if for each* $x \in X$ $b \leq x$. *The* greatest lower bound *or infimum of* X *denoted by* $infX$ $(\wedge X)$ *is the greatest of all lower bound of* X.

Definition 2.5 *A* lattice $\mathcal{L}$ *is a poset* P *such that any two elements* $x, y \in P$ *have a supremum called* 'join' *and an infimum called* 'meet'. *Therefore join and meet are (total, binary) operations on a poset* P *and for any two elements* x *and* y *of a lattice we write* $x \vee y$ *for their join and* $x \wedge y$ *for their meet.*

[1] For a detailed discussion see Birkhoff (1967)

[2] Notice that $\leq$ is used here in a more general sense than in the case of concepts.

Definition 2.6 *A lattice $\mathcal{L}$ is called* complete *if each subset $X \subseteq P$ has a supremum and an infimum in $\mathcal{L}$. We also say that $\mathcal{L}$ is closed under supremum and infimum.*

Since we restricted ourselves to finite examples, all our lattices will be also finite. Since any finite lattice is complete, we can take completeness of our lattices for granted.

Definition 2.7 *A subset D of a complete lattice $\mathcal{L}$ is called* infimum dense *if each element of the a lattice is an infimum of a subset X of D i.e. $\mathcal{L} = \{\wedge X \mid X \subseteq D\}$. Similarly a subset of a complete lattice $\mathcal{L}$ is called* supremum dense *if $\mathcal{L} = \{\vee X \mid X \subseteq D\}$.*

Now we can state the basic theorem mentioned earlier (Wille, 1992):

Theorem 2.1 *$\underline{\mathbf{L}}(O, P, \gamma)$ is a complete lattice of the context (O, P, γ) whose infimum is:*

$$\bigwedge_{t \in T}(A_t, B_t) = (\bigcap_{t \in T} A_t, (\bigcup_{t \in T} B_t)'') \tag{9}$$

and supremum is:

$$\bigvee_{t \in T}(A_t, B_t) = ((\bigcup_{t \in T} A_t)'', \bigcap_{t \in T} B_t \tag{10}$$

Conversely, a complete lattice $\mathcal{L}$ is isomorphic to $\underline{\mathbf{L}}(O, P, \gamma)$ if and only if there are mappings $\sigma : O \longrightarrow \mathcal{L}$ and $\varrho : P \longrightarrow \mathcal{L}$ such that σO is supremum–dense in $\mathcal{L}$ and ϱP is infimum–dense in $\mathcal{L}$, and $o \gamma p$ is equivalent to $\sigma o \leq \varrho p$ for all $o \in O$ and $p \in P$, i.e. $\mathcal{L} \cong \underline{\mathbf{L}}(\mathcal{L}, \mathcal{L}, \leq)$ and, if $\mathcal{L}$ has a finite length, $\mathcal{L} \cong \underline{\mathbf{L}}(\mathcal{J}(\mathcal{L}), \mathcal{M}(\mathcal{L}), \leq)$.
We call $\underline{\mathbf{L}}(O, P, \gamma)$ a concept lattice.

3. Generating and Drawing Concept Lattices

Before going into the discussion of how to interpret information provided by concept lattices or rather how to use concept lattices for design purposes, let us discuss the following questions: How can we obtain the concept lattice of a given context, and how can the concept lattice be represented conveniently by easily surveyable line diagrams? In fact, the primary aim of this paper is to discuss the use of such diagrams as decision support tools in design rather than discussing theoretical questions related to concept lattices.

The most obvious derivation method to obtain all concepts from a context (O, P, γ) is to generate for all subsets $X \subseteq O$ the concept (X'', X') by 5 or for all subsets $Y \subseteq P$ the concept (Y', Y'') by 6 By ordering the concepts using the $\leq$–relation, we obtain the concept lattice $\underline{\mathbf{L}}(O, P, \gamma)$. Even though this method might work well for smaller contexts, it requires the generation of at least $2^{|O|}$ subsets and their "prime" derivates, if $O \leq P$ or at least $2^{|P|}$ subsets if $P < O$. (It is also possible to use a combination of subsets X and Y in order to obtain all concepts.) In case of larger contexts, this method becomes obviously inefficient. It is more

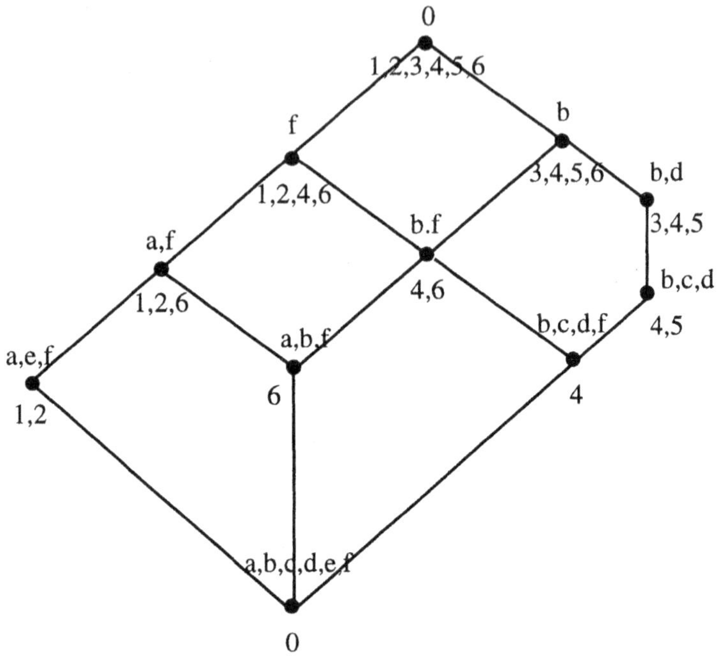

Figure 3. Hasse diagram of the concept lattice generated from the building structure context.

efficient to use the formulae $X' = \bigcap_{o \in X}\{o\}'$ or $Y' = \bigcap_{p \in Y}\{p\}'$ which are special cases of 7 and 8, and then form (X'', X') or (Y', Y''). This means that we can start with arbitrary intents $\{o\}'(o \in O)$ or extents $\{p\}'(p \in P)$ and derive new extents and intents by the above formulae. Repeating the process provides all concepts, since every extent or intent is the intersection of certain other extents or intents. There are other, more efficient ways to generate the concepts by computer as well. For a discussion and the comparison of concept generating algorithms see Gantner (1987).

Lattices, in particular concept lattices, are usually represented by line diagrams, called Hasse diagrams. The nodes of the diagrams represent concepts and if we have $(A_1, B_1) \geq (A_2, B_2)$ then we place the node representing (A_1, B_1) higher in the diagram than the node representing (A_2, B_2). If $(A_1, B_1) \geq (A_2, B_2)$ and there is no concept (A_3, B_3) such that $(A_1, B_1) \geq (A_3, B_3) \geq (A_2, B_2)$ then the nodes representing (A_1, B_1) and (A_2, B_2) are connected by a line in the diagram. This representation of the concept lattice is a graph displaying the ordering relation. Such a graph is called Hasse diagram. Figure 3 shows the Hasse diagram of the concept lattice generated from the context of Figure 1. In the diagram the intent is placed above and the extent below the corresponding node.

It is, however, sufficient to attach the name of an object o to a node representing $\sigma o := (\{o\}'', \{o\}')$ which is the *smallest* concept containing o in its extent, and

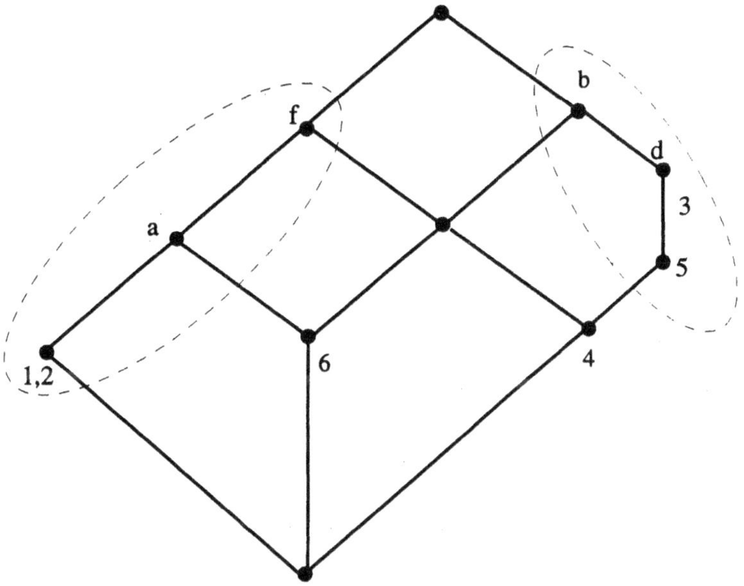

Figure 4. The simplified Hasse diagram of the building structure lattice.

similarly, it is sufficient to attach the name of a property to the node representing $\varrho p := (\{p\}', \{p\}'')$ which is the *largest* concept containing p in its intent. This follows from $o \in A \Leftrightarrow \sigma o \leq (A, B)$ and $p \in B \Leftrightarrow \varrho p \geq (A, B)$. Figure 4 shows the simplified Hasse diagram. The missing labels can easily be obtained by consecutive addition of extents or intents along the upward or downward sloping paths.

Any concept (or other kinds of) lattice can be represented by an infinite number of different Hasse diagrams. Some diagrams are more, other less appropriate for representing and analyzing lattice properties. Therefore, regardless of the particular use of the lattice, there are some guide lines for generating 'nice' and useful diagrams. The most important one is that the diagram should display or even emphasize important structural lattice properties. This can be achieved by various methods and we will discuss one of these methods in section 5. It is also desirable that connecting lines are straight sloping segments. [3] It is also common to draw planar or 'as planar as possible' diagrams, i.e., to minimize the number of crossing lines. Hasse diagrams are often composed of parallelograms or rather the underlying grid of the diagram is parallelogram grid. Boolean lattices can nicely be represented by n–dimensional cubes drawn on such grids. This is the reason that Hasse diagrams are often composed of Boolean sublattices depicted as n–dimensional cubes. (For

[3] Sometimes, however, we want to represent important features that can be achieved only by abandoning straight lines.

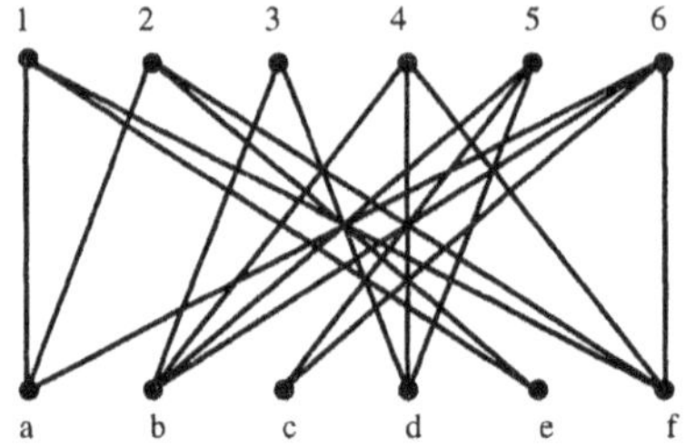

Figure 5. Representation of the building structure context by a bipartite graph.

more on the representation of lattices by line diagrams see Rival (1989).)

Besides concept lattices, other methods can be used for representing structural properties of a context as well. First of all the context might not be given in the (convenient) matrix form as in Figure 1, such that recognizing objects, properties and relations itself can be a representational task. For example, the list $\ell_P = (1, 2, 6; 3, 4, 5, 6; 4, 5; 3, 4, 5; 1; 2; 1, 2, 4, 6)$, or the list $\ell_O = (a, e, f; a, e, f; b, d; b, c, d, f; b, c, d; a, b, f)$ describes the exact same context as Figure 1. Although this representation is more efficient and better for programming purposes, it is hard to recognize structural properties by such lists. We can also represent the same context by a bipartite graph as it is shown in Figure 5. This representation is also hard to read and provides little information about the structure of the context.

In contrast, as Wille (1992) states: "Concept lattices constitute a structural analysis of data contexts without reducing the the data. A labeled line diagram of a concept lattice still *represents all knowledge*[4] coded in the underlying context and, furthermore, unfolds (and reveals to the eye) the inherent conceptual structure of the coded knowledge." In fact, concept lattices display information provided by all other structural representation methods. For example *hierarchical concept clustering* provides clusters like the one of building structures with good fire and sound proof capabilities, or the one of those with short construction time and low weight (see Figure 6). The same clusters can easily be obtained from the concept lattice in Figure 4 where we can find these two clusters on the left and right branches of the diagram respectively.

The concept lattice also represents *all implications* between the attributes. An implication $Y \longrightarrow Z$ of a context (O, P, γ) is a pair of subsets $Y, Z \subseteq P$ such that $Y' \subseteq Z'$, i.e., each object of O that has all the properties of Y will also have the properties of Z. Similarly, the implication $X \longrightarrow T$ is a pair of subsets $X, T \subseteq O$ such that $X' \subseteq T'$,i.e., each property of P that belongs to the the objects of X will belong also to the objects of T. This can be considered as an 'inheritance'–relation in object oriented terms that is displayed by the diagram. We can read the implications from the diagram because of Theorem 2.1, $Y' \subseteq Z' \Leftrightarrow \forall l \in Z$:

[4]Emphasis added.

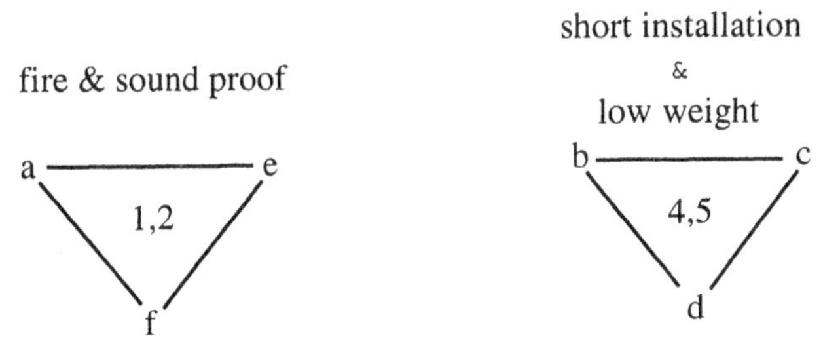

Figure 6. Two clusters whose elements share the same properties.

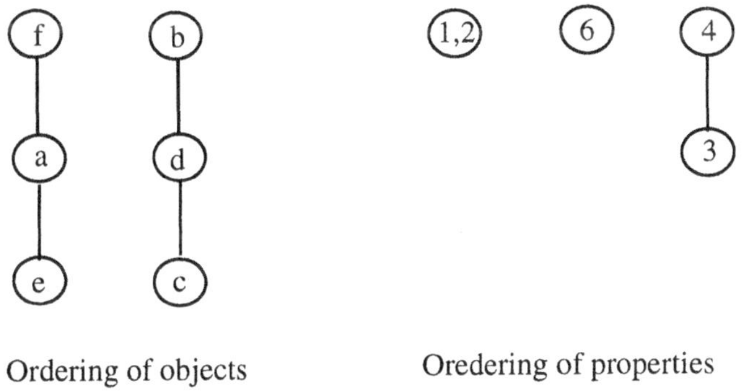

Figure 7. Ordering of the properties and objects.

$(Y', Y'') \leq \varrho l$ in $\underline{\mathbf{L}}(O, P, \gamma)$ and $(Y', Y'') = \bigwedge_{k \in Y} \varrho k$ and $X' \subseteq T' \Leftrightarrow \forall n \in T : (X'', X') \leq \sigma n$ in $\underline{\mathbf{L}}(O, P, \gamma)$ and $(X'', X') = \bigvee_{m \in X} \sigma m$. For example $\{a, c\} \longrightarrow \{e, f\}$ because $\sigma a \wedge \sigma c \leq \sigma e$ and $\sigma a \wedge \sigma c \leq \sigma f$, i.e., the properties of a and c will also be properties of e and f.

Implications with one element premise provide a natural ordering on the set of properties by $k \leq l :\Leftrightarrow l \longrightarrow k (\Leftrightarrow \varrho l \leq \varrho k)$ and similarly, we obtain a natural ordering on the set of objects by $m \leq n :\Leftrightarrow m \longrightarrow n (\Leftrightarrow \sigma m \leq \sigma n)$. This ordering is shown in Figure 7. Figure 8 shows implications on the objects.

4. Analyzing Contexts by Concept Lattices

The following example from Lenart (1990) describes a common design task by statements and their connections. (Another example in Lenart (1988) shows another novel application of concept analysis in design.) The extents of the concepts are statements about the current situation at a certain university describing the problem of having unsatisfactory student accommodations. Using Ho's (1982a) terms, these statements represent the *state space* of the design process. The intents of the concepts are sets of decisions changing the current situation. This corresponds to the *decision space* (Ho, 1982a). The context is represented as a matrix in Figure

f →	a,e	b,c →	d
a →	e	b,d →	c
b →	d,c	b,e →	a,c,d,f
d →	c	b,f →	a,c,d,e
a,b →	c,d,e,f	a,b,c →	d,e,f
a,c →	e,f	a,b,d →	c,e,f
a,d →	b,c,e,f	a,b,e →	c,d,f
a,f →	e	a,b,f →	c,d,e

Figure 8. Implications on the objects.

9. In this context $x\gamma y$ means that the statement x is effected by the decision y, or in other words, the decision y changes the validity of the statement x. We notice that we did not specify *how* the decision y effects the statement x nor the strong or weakness of the influence. However, by refining the relation between the two sets (i.e. expanding the context) several modifications are possible. The new context replaces a single statement by a certain number of sub statements and defines the relation on these new statements. Such a method for the extension or refinement of contexts was described in Macgill (1983). However, it is also possible to have many values or a scale describing the relationship between objects and properties. A method for dealing with many-valued contexts is described in Wille (1987; 1992).

The problem is obviously simplified by describing it with only a few number of statements using a binary relation whether certain statements are mutually related or not. Although this fictive problem is quite common and realistic, this small example shows that even in such simplified cases the context can be quite complicated, yet concept analysis provides a powerful tool to analyze the problem and to help in the decision making process. Like in the previous example, the concept lattice can be generated by calculating all possible concepts, ordering the concepts, and finally drawing the concept lattice. Even though the context of this simple task is small, it provides 63 concepts and the resulting lattice, as Figure 10 shows, is quite complex.

Obviously, there are infinitely many different drawings representing the same lattice. However, not all the diagrams represent the lattice 'nicely'. A 'nice' drawing means that we can 'read' the diagram, i.e. obtain information (dependencies) easily. One trick we might want to use is to look for Boolean sublattices (or 'almost'–Boolean sublattices, as mentioned earlier) and draw them as n–dimensional cubes with parallel edges. The search for Boolean sublattices is not difficult and can systematically be done. However, putting the entire diagram together requires some skill. Because of the difficulties in drawing Hasse–diagrams (especially 'nice' ones)

problems \ decisions	Building new dormitories on the university's building site	Improve the university's transportation system	The university signs long term contracts with appartment owners close to campus	Giving preferences to and raising the number of students living in the university area	The university contributes to the students' traffic expences	Supporting students for renting appartments by university allowences	Reduce the admission number of new students	Building new parking facilities on campus	Reduce the university profile (soem departments will be abandoned)	
The number of students living in dorms close to the campus is limited	x		x	x	x				x	1
Renting an apartment close to the campus is expensive	x	x	x	x	x	x	x		x	2
No new dormitory can be built on campus	x	x	x			x			x	3
The university has a building site outside the city in driving distance	x						x	x		4
The university has resources (annual budget) for unspecified building activities	x							x		5
The university has resources for supporting a limited number of students			x		x	x	x			6
There are government resources for building dormitories and parking facilities	x							x		7
There are government resources for unspecified/unrestricted student support			x		x	x	x		x	8
The university has not enough parking facilities	x	x		x	x	x	x	x	x	9
The university has not enough busses for transporting students	x	x		x	x	x	x	x		10
There is a small number of students commuting between home and the univ.	x	x	x	x	x		x	x	x	11
The university's teaching capacity is overloaded	x		x	x	x	x	x	x	x	12
	a	b	c	d	e	f	g	h	i	

Figure 9. The context of the student accommodation problem.

and because of this is a time consuming process, there is an ongoing research effort to generate lattices automatically by computer programs (Wille, 1989). This is important because the concept lattice describing a given task changes with each new data or modification of the context. Frequent — hypothetic or concrete — changes can only be analyzed if one can generate Hasse–diagrams quickly by computers.

Analyzing a Hasse–diagram that represents a design context, means the search for concepts providing satisfactory solutions. Since what 'satisfactory solution' means is subjective, the diagram itself doesn't provide a solution. It supports, however, the decision making process by displaying all the concepts and their order-

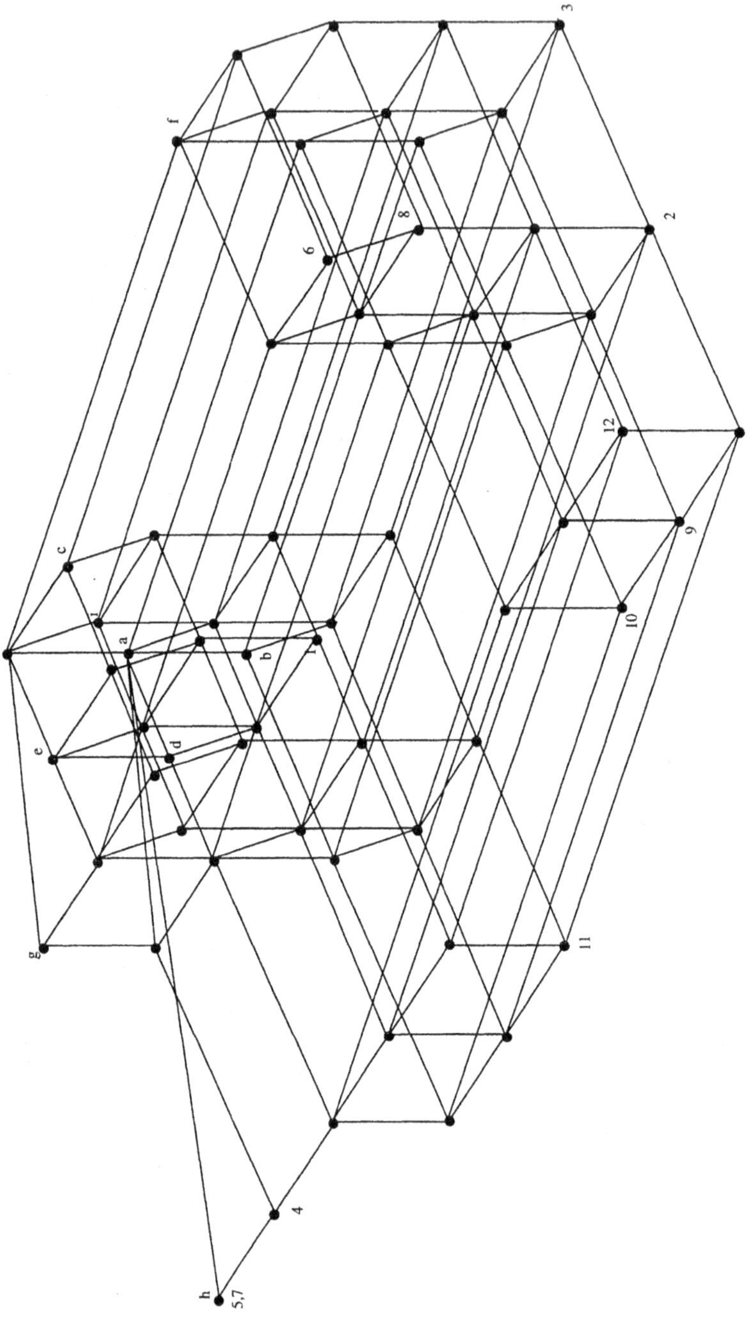

Figure 10. The concept lattice of the student accommodation problem.

ing. For instance, if we choose a set of decisions of a concept C_i, let's say $\{a, b, i\}$ then it changes the state of the 'objects' $\{2, 3, 9, 11\}$ of the same concept. In other words, the decisions a, b and i (may) solve the problems 2, 3, 9 and 11. Any set of decisions of a concept C_j that is connected by a continuous, steadily downward sloping path to the concept C_i solves (or effects) just a (proper) subset of these tasks.

Or we can do the reasoning in the other direction by choosing a set of 'objects' that are essential for solving this particular problem. Similarly, any concept in our lattice containing the chosen objects has a set of decision that will effect these objects. Moreover, any continuous, upward sloping path starting at such a concept leads to concepts with the same property.

In both cases; starting the process with decisions or objects, we might choose a set that does not occur in the lattice. Thus, if we felt that the set 3, 9, 11 would have been the key to our context then we would look at the concepts containing these elements. The concept lattice tells us that there is no such set of decisions in our context that effects exactly these three objects and nothing else. The lattice also tells us that the smallest such concept is $A = \{2, 3, 9, 11\}, B = \{a, b, i\}$, and there are exactly five other concepts on up–sloping paths starting at this concept which have the same property.

The next step of the process is to decide which of these six concepts is the most satisfactory. This is a kind of optimization where we have to compare different sets of decisions, or rather their consequences. It is likely that the process involves both: searching for an 'optimal' design state and also for 'optimal' decisions. Optimality means that we want to improve the current situation as much as possible with the minimum amount of effort and sacrifice. It is obvious that design in general is far from being optimizable, and because of the complexity of design problems, it is important that concept lattices (in particular Hasse–diagrams) help us to describe design tasks in exact terms. This description, or model enables us to compare and evaluate decisions. In case of team or collaborative design, concept lattices provide a powerful argumentative tool for discussions and negotiations.

5. Generating Lattices by Subdirect Products

Although Hasse diagrams can help us to display and analyze complex design tasks, the generation of Hasse–diagrams becomes increasingly difficult — if not impossible — by the increase of the number of concepts. With increasing number of concepts the diagram becomes large, dense and cluttered and at some point it might fail to provide any help. Not just generating but also 'reading' such diagrams can be difficult. Additionally, not all of the hierarchical structures hidden in the original context are equally significant for the design process. Hasse–diagrams so far do not allow us to make distinctions between different dependencies.

All these problems can be resolved by means of suitable sublattices. Before going into the discussion of sublattices, however, let us look at again the Hasse–diagram in Figure 10. It is not difficult to discover the Boolean sublattices in the diagram having parallel edges. Moreover, we can easily discover that the entire diagram contains two similar looking sublattices (one in the left upper part and the other in the right bottom part of the picture) that are connected by a bundle of parallel lines. By removing these lines the diagram becomes simpler and we still have all the informations of the original diagram. From the simplified diagram in Figure 11 we can obtain the original diagram by moving the 'boxed' diagrams of one sublattice over the diagram of the other sublattice along the connection line of the two boxes. Two points are connected in the original diagram if they cover each other following this translation. Our claim is that we can apply this 'trick' to any concept lattice, and not only we will have a way to represent large lattices efficiently but also we will be able to organize the diagram so that certain structural dependencies become transparent.

As the above described intuitive simplification indicates, sublattices are the key to representation problems. In fact, we are seeking methods to generate and represent lattices from smaller sublattices. There are a couple of methods for generating lattices from appropriate sublattices or decomposing concept lattices into sublattices (Wille, 1983; Wille, 1985). Here we show how lattices can be obtained as the subdirect product of sublattices (Wille, 1987).

Let us first consider contexts having certain structural properties. The aim is to have special cases in which the concept lattice can be generated easily by suitable subcontexts. In other words the idea is to find structural properties of a context which allow us to generate the concept lattice from sublattices. For this purpose, we introduce the following notations: Let us denote a lattice $\mathcal{A}$ by $(X, \leq_{\mathcal{A}})$, where X is the underlying set of $\mathcal{A}$ and $\leq_{\mathcal{A}}$ its partial order.

Definition 5.1 *The* direct product $\mathcal{C} = \mathcal{A} \times \mathcal{B} = (X \times Y, \leq_{\mathcal{C}})$ *of two lattices* $\mathcal{A} = (X, \leq_{\mathcal{A}})$ *and* $\mathcal{B} = (X, \leq_{\mathcal{B}})$ *is defined such that* $(x_1, y_1) \leq_{\mathcal{C}} (x_2, y_2)$ *if and only if* $x_1 \leq_{\mathcal{A}} x_2$ *and* $y_1 \leq_{\mathcal{B}} y_2$*. A bounded lattice is a lattice with greatest and smallest element.*

Definition 5.2 *The* horizontal sum $\mathcal{A} \oplus_h \mathcal{B}$ *of two bounded lattices* $\mathcal{A} = (X, \leq_{\mathcal{A}})$ *and* $\mathcal{B} = (Y, \leq_{\mathcal{B}})$ *is obtained from their cardinal sum* $(X \dot{\cup} Y, \leq_{\mathcal{A}} \dot{\cup} \leq_{\mathcal{B}})$ *(where* $\dot{\cup}$ *denotes disjoint union) by identifying the smallest and the greatest element of the two bounded lattices respectively.*

Definition 5.3 *The* vertical sum $\mathcal{A} \oplus_V \mathcal{B}$ *is obtained from their ordinal sum* $(X \dot{\cup} Y, \leq_{\mathcal{A}} \dot{\cup} \leq_{\mathcal{B}} \cup (X \times Y))$ *by identifying the greatest element of* $\mathcal{A}$ *with the smallest element of* $\mathcal{B}$.

Definition 5.4 *A* subdirect product $\mathcal{D} = (S, \leq_{\mathcal{D}})$ *of the lattices* $\mathcal{A} = (X, \leq_{\mathcal{A}})$ *and* $\mathcal{B} = (X, \leq_{\mathcal{B}})$ *is a subset* $S \subseteq X \times Y$ *closed under the operations* $\vee$ *and* $\wedge$ *(sublattice) such that for each element* $x \in X$ *(or* $y \in Y$*) there is an element* $s \in S$ *having* x *(or* y*) as its component.*

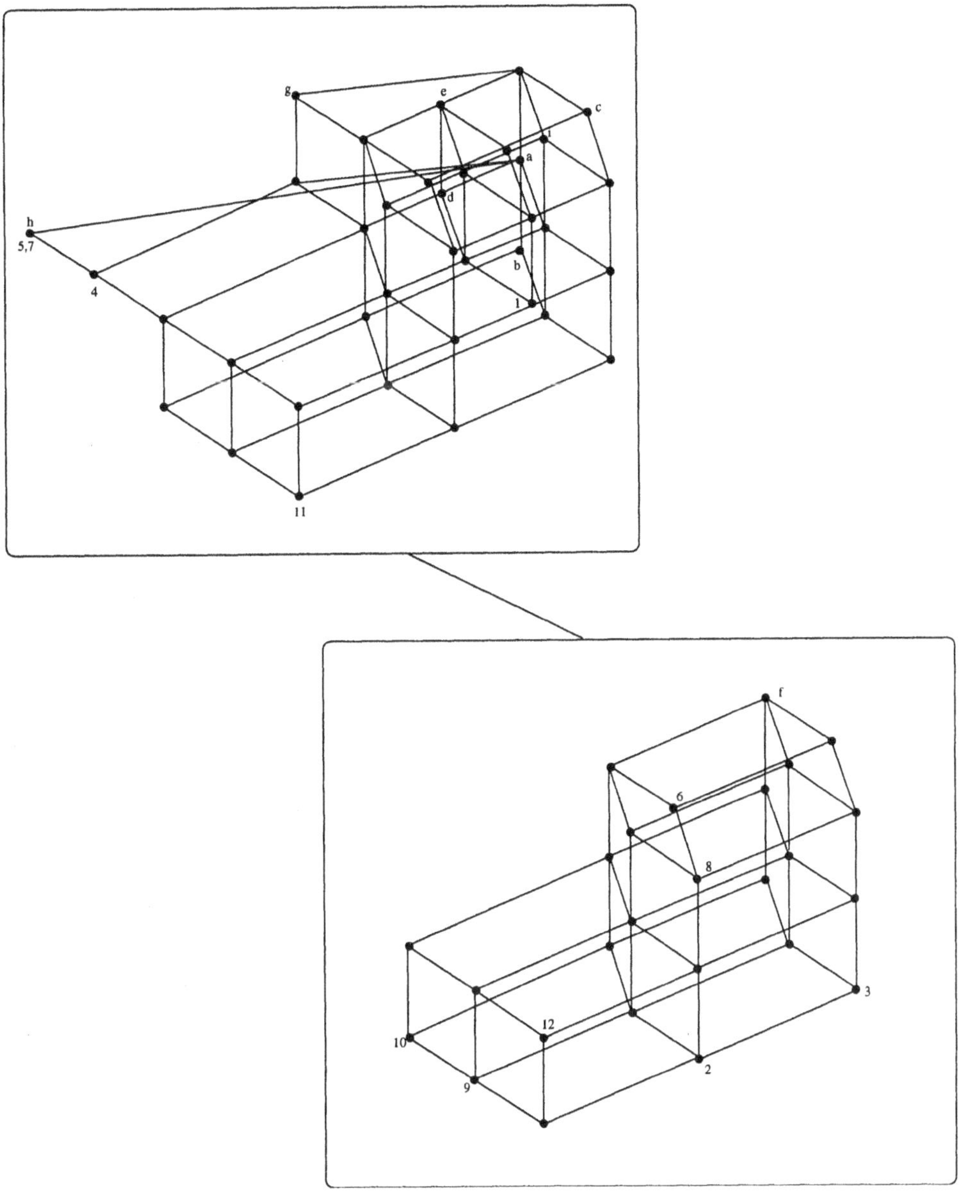

Figure 11. Simplifying the diagram of the lattice describing the student accommodation problem.

The following theorem is applicable in cases where the context contains disjoint subcontexts whose polar sets are empty (Wille, 1987).

Theorem 5.1 *Let* $C_1 = (X_1, Y_1, \gamma_1)$ *and* $C_2 = (X_2, Y_2, \gamma_2)$ *be contexts with* $X_1 \cap X_2 = \emptyset, Y_1 \cap Y_2 = \emptyset, X_1^* = X_2^* = Y_1^* = Y_2^* = \emptyset$ *and* $\mathcal{L}_{C_1} = \mathcal{L}_1(X_1, Y_1, \gamma_1)$ *and*

$\mathcal{L}_{C_2} = \mathcal{L}_2(X_2, Y_2, \gamma_2)$ *the lattices belonging to them.*

Then $\mathcal{L}_C(X_1 \cap X_2, Y_1 \cap Y_2, \gamma_1 \cap \gamma_2)$ *is isomorphic to the* horizontal *sum of* $\mathcal{L}_1$, *and* $\mathcal{L}_2$, $\mathcal{L}_C(X_1 \cap X_2, Y_1 \cap Y_2, \gamma_1 \cap \gamma_2 \cap X_1 \cap Y_2)$ *is isomorphic to the* vertical *sum of* $\mathcal{L}_1$, *and* $\mathcal{L}_2$, $\mathcal{L}_C(X_1 \cap X_2, Y_1 \cap Y_2, \gamma_1 \cap \gamma_2 \cap (X_1 \times Y_2) \cap (X_2 \times Y_1))$ *is isomorphic to the* direct product *of* $\mathcal{L}_1$ *and* $\mathcal{L}_2$.

If Theorem 5.1 can not be applied, i.e. we do not have the properties allowing the application of this theorem, then a larger concept lattice can be still generated by using the following general theorem:

Theorem 5.2 *Let* $C = (X, Y, \gamma)$ *be a context, let* $\{X_i \mid i \in I\}$ *be a partition of* X *and let* $\{Y_j \mid j \in J\}$ *be a partition of* Y. *Then a* $\vee$*–embedding of* $\mathcal{L}_C(X, Y, \gamma)$ *into the direct product of the* $\mathcal{L}_C(X_i, Y, \gamma \cap (X_i \times Y))_{i \in I}$ *is given by* $\langle A, B\rangle \mathcal{P} \langle A \cap X_i, (A \cap X_i^*)_{i \in I}\rangle$, *a* $\wedge$*–embedding of* $\mathcal{L}_C(X, Y, \gamma)$ *into the direct product of the* $\mathcal{L}_C(X, Y_j, \gamma \cap (X \times Y_j))_{j \in J}$ *is given by the mapping* $\langle A, B\rangle \mathcal{P} \langle B \cap Y_j, (B \cap Y_j^*)_{j \in J}\rangle$.

Conversely, each concept lattice can be represented as a subdirect product of sublattices. We obtain sublattices by dissecting the set Y (or X) of a context $C = (X, Y, \gamma)$ into subsets $Y_1, \cup Y_2 \cup \ldots \cup Y_n = Y$ (or $X_1, \cup X_2 \cup \ldots \cup X_n = X$) and generating the lattice $\mathcal{L}_{C_j}(X, Y_j, \gamma \cap (X \times Y_j))_{j \in J}$ (or $\mathcal{L}_{C_i}(X_i, Y, \gamma) \cap (X_i \times Y)_{i \in I}$).

Let us consider the simplest case of a dissection having $Y = Y_1 \cup Y_2$ of the context $C = (X, Y, \gamma)$. In this case we obtain the two concepts $C_1 = (X, Y_1, \gamma_1)$, where $\gamma_1 = \gamma \cap (X \times Y_1)$ and $C_2 = (X, Y_2, \gamma_2)$ where $\gamma_2 = \gamma \cap (X \times Y_2)$. From the concepts C_1 and C_2 we obtain the two sublattices $\mathcal{L}_{C_1}(X, Y_1, \gamma_1)$ and $\mathcal{L}_{C_2}(X, Y_2, \gamma_2)$. The representation of the lattice $\mathcal{L}_C(X, Y, \gamma)$ as the subdirect product of $\mathcal{L}_{C_1}$ and $\mathcal{L}_{C_2}$ goes as follows:

We draw the lattice $\mathcal{L}_{C_2}$ as a diagram having 'boxes' (rectangles) instead of points. In each box we draw the Hasse–diagram of the lattice $\mathcal{L}_{C_1}$ identically such that moving one box to another along the edges the pictures of $\mathcal{L}_{C_1}$ cover each other (i.e. the diagrams are congruent). Let $\{x\}^* \cap Y_1 = A$ and $\{x\}^* \cap Y_2 = B$. Now the concept $\langle A, B\rangle$ of the lattice $\mathcal{L}_C$ is in the box $\langle B^*, B\rangle$ of the diagram of $\mathcal{L}_{C_2}$ and at the point $\langle A^*, A\rangle$ of the diagram of $\mathcal{L}_{C_1}$. Usually, some parts of the diagrams of the lattices $\mathcal{L}_{C_1}$ are omitted since not all of the depicted concepts exist. The resulting diagram is a unique representation of $\mathcal{L}_C$, i.e. from the 'box diagram' of the lattice one can easily reconstruct the lattice itself. However, it contains less edges than the Hasse–digram of the lattice since 'parallel' and 'complete' edges are replaced by single ones. Parallel edges connect two diagrams in adjacent boxes such that each pair of connected points would cover each other by moving the congruent diagrams. Complete edges connect a single (top or bottom element of the diagram with all the points of a diagram in an adjacency box. (Because of the transitivity of $\leq$ and $\geq$ we need to connect just the smallest or greatest elements in the box.)

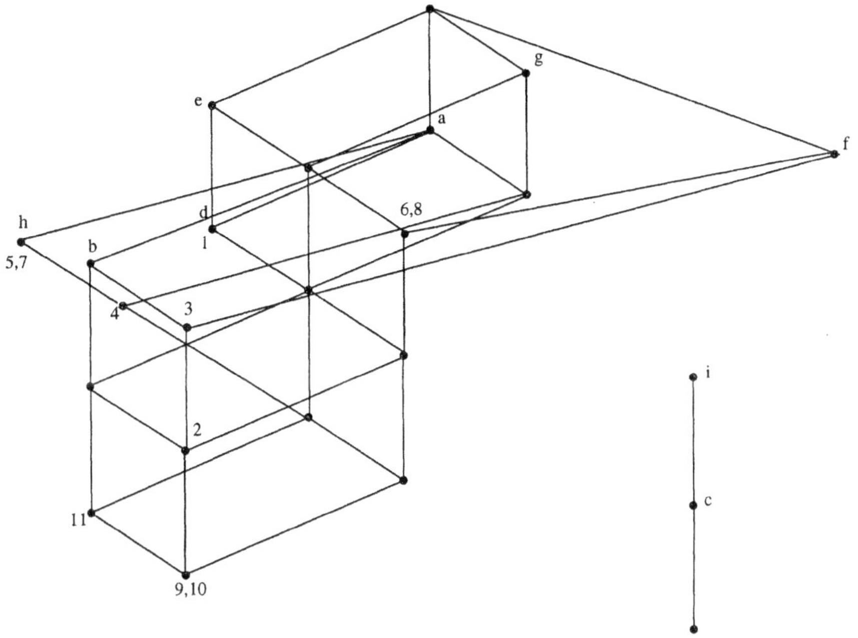

Figure 12. The two sublattices generating the box diagram in Figure 11

The dissection of the set $Y = \{a, b, c, d, e, f, g, h, i\}$ of our example into the subsets $Y_1 = \{a, b, c, d, e, g, h, i\}$ and $Y_2 = \{f\}$ provides the two sublattices on Figure 12. The product of these sublattices provides the lattice on Figure 11.

Another dissection of Y into $Y_1 = \{a, b, c, d, e\}$ and $Y_2 = \{f, g, h, i\}$ provides interestingly two very similar sublattices. Figure 13 shows both: the two sublattices $\mathcal{L}_{C_1}(X, Y_1, \gamma_1)$ and $\mathcal{L}_{C_2}(X, Y_2, \gamma_2)$ and the box diagram of the lattice $\mathcal{L}_C(X, Y, \gamma)$.

If the dissection is based on some structural properties then the box diagram displays the order structure according to these structural properties. We could for instance divide the statements about the current planning situation into financial statements $X_1 = \{2, 5, 6, 7, 8\}$ and object descriptive statements $X_2 = \{1, 3, 4, 9, 10, 11, 12\}$. In this case the box diagram would display dependencies that are financial and non-financial nature by having two different representation levels.

By having a decomposition of the set Y (or X) into three, four, etc. subsets the depth of the structure grows. It becomes a three, four, etc. level box diagram. Even very large concept lattices become representable by using such higher level, nested structures. Besides the fact that nested, multilevel diagrams help us to reduce the number of edges further and make the diagram survey-able, they also provide us more structuring possibilities. In nested diagrams each level represents an aspect we want to emphasize or make transparent.

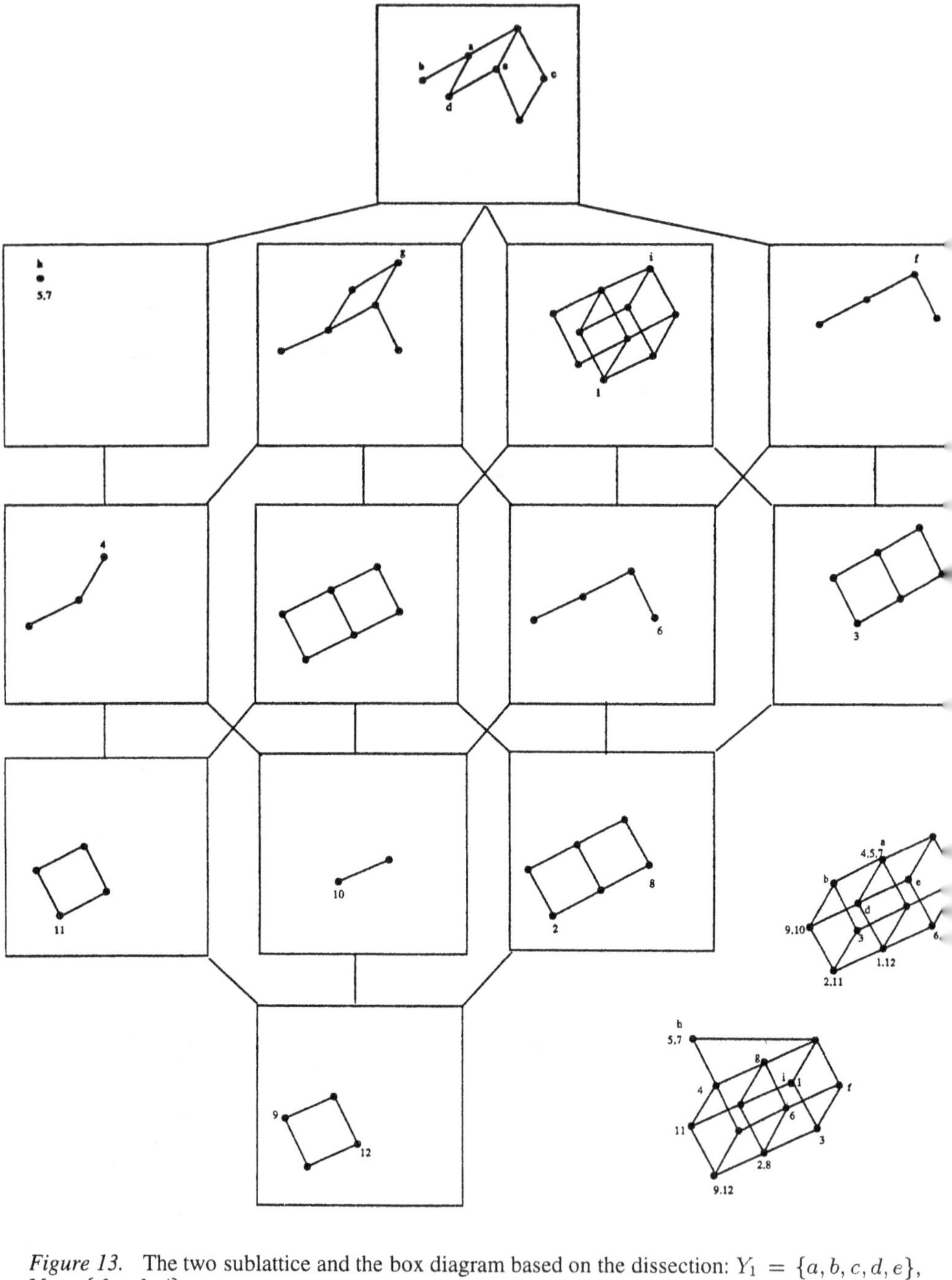

Figure 13. The two sublattice and the box diagram based on the dissection: $Y_1 = \{a, b, c, d, e\}$, $Y_2 = \{f, g, h, i\}$

6. Conclusion

The paper presents the fundamentals of concept analysis and how it can be used for analyzing design tasks. In particular, it shows how to apply Wille's analysis and representation method in design. By using results in lattice theory, we can generate, analyze concept lattices and draw their Hasse–diagrams. Since Hasse–diagrams provide easy access to and easy understanding of concept lattices (even for a layman), the focus of the paper lies in the development and representation of Hasse–diagrams. Hasse diagrams provide an insight of complex relations that are hidden in the context and usually difficult to obtain otherwise. Since design contexts have not been analyzed so far by these methods, there is a wide range of new applications for concept lattices in the design field.

References

Atkin, R. H.: 1974, *Mathematical Strcture in Human Affairs*, Heinemann, London.

Birkhoff, G.: 1967, *Lattice Theory* (3rd edn), Mathematical Society, Providence, RI.

Gantner, B.: 1987, Algorithmen zur formalen begriffsanalyse, *in* K. E. Wolf, B. Ganter and R. Wille (eds), *Betraege zur Begriffsanalyse*, B.I. Wissenschaftsverlag, Mannheim.

Ho, Y-S.: 1982a, The planning process: fundamental issues, *Environment and Planning B*, **9**, 387–395.

Ho, Y-S.: 1982b, The planning process: structure of verbal descriptions, *Environment and Planning B*, **9**, 397–420.

Ho, Y-S: 1982c, The planning process: a formal model, *Environment and Planning B*, **9**, 377–386.

Lenart, M.: 1988, Axiomatic approach to analyzing floor plan topology, *4th International Conference on System Research, Informatics and Cybernetics*, Baden-Baden, Germany, pp. 63–71.

Lenart, M.: 1990, Concept lattices as planning models, *Design Methods and Theories*, **24**(1), 1136–1163.

Macgill, S. M.: 1983, A consideration of Johnson's Q-discrimination analysis, *Environment and Planning B*, **9**, 315–330.

March, L.: 1982, On Ho's methodological approach to design and planning, *Environment and Planning B*, **9**, 421–427.

Rival, I.: 1989, Graphical data structures for ordered sets, *in* I. Rival (ed.), *Algorithms and Order*, Kluwer, Dordrecht, pp. 3–31.

Wille, R.: 1981, Restructuring lattice theory: An approach based on hierarchies of concepts, *in* I. Rival (ed.), *Proceedings of the Symposium on Ordered Sets*, Banff, pp. 445–470.

Wille, R.: 1983, Subdirect decomposition of concept lattices, *Algebra Universalis*, **17**(3), 275–283.

Wille, R: 1984, Line diagrams of concept systems, *International Classification*, **11**(2), 77–86.

Wille, R.: 1985, Tensorial decomposition of concept lattices, *Order*, **2**, 81–95.

Wille, R.: 1987, Subdirect product construction of concept lattices, *Discrete Mathematics*, **63**, 305–313.

Wille, R.: 1989, Lattices in data analysis: How to draw them with a computer, *in* I. Rival (ed.), *Algorithms and Order*, Kluwer, Dordrecht, pp. 33–58.

Wille, R.: 1992, Concept lattices and conceptual knowledge systems, *Computers Math. Applic.*, **23**(6-9), 493–515.

12

SUPPORTING THE DESIGN PROCESS BY AN INTEGRATED KNOWLEDGE BASED DESIGN SYSTEM

HANS GRABOWSKI, RALF-STEFAN LOSSACK AND CLEMENS WEIS
University of Karlsruhe, Germany

Abstract. The German design methodology demonstrated its usefulness for solving design problems by being applied in the enterprises' design departments during the years. Design methodologists as Roth, Pahl, Beitz and Hubka developed an instrument for a methodological approach to design tasks. This approach is a strongly process oriented one and describes together with the fundamentals of design, general strategies for solving design problems. Another approach, which was followed by the researchers of design systems was an information oriented one. Here the main work was concentrated on the modeling of the information needed in design. There was no approach which combined consequently these two different approaches. In this article we introduce in the first part the fundamental aspects of the German design methodology by describing the modeling space of design with the help of an example of mechanical engineering. In the second part we introduce design working spaces which help to structure and administer design solutions. Finally an approach to incorporate the process oriented aspect of design into a knowledge based CAD system is presented.

1. Introduction and Overview of the Design Process

Design methodology considers the design process in an idealized manner as a successive concretion of the description of the to-be characteristics of a technical object (Koller, 1985; Pahl and Beitz, 1994; Roth, 1994 and others). This concretion process takes place on the product modeling level. Koller, Pahl, Beitz, Roth and other design methodologists define this process to lead from

- The incomplete to the complete,
- The abstract to the concrete,
- The rough to the precise,
- The provisional to the definitive and
- Possible alternatives to the optimal solution.

An important characteristic of this process is the successive growth of the set of design characteristics with respect to the current state of the design. Here design characteristics are defined as the instantiated solution properties of a product to be developed. The sum of all these solution properties of a product characterises in connection with the corresponding product model the properties and the overall behaviour of the product in real life.

The description of the design characteristics mentioned above is the result of the design process. They can be assigned to the design phases known from the design methodology. The design phases consist of defining the requirements of a product, of the definition of the functional structure and the function flow within a product, the description of the physical effects which can be assigned to the respective functions in correspondence with the modeling of the product's effective geometry and the design of the product's shape. From the methodological point of view of design the origin for the phase orientation of the design process is found here. In accordance to these phases the logical modeling layers of a design system have been defined as follows:

1.1. REQUIREMENTS MODELLING LAYER

The requirements modeling layer serves for the computational projection of the results won by the clarification of the design task. This contains the preconditions of the design, the definition of product requirements, i.e. the to-be properties of the future product and the description of the product's immanent task structure as the transition to the functional modeling.

1.2. FUNCTIONAL MODELLING LAYER

Functional modeling serves for finding and describing the functions of a design solution to be developed, as well as the functional interrelationships within the future product. Functional modeling allows the definition and manipulation of functions on different levels of abstraction, as well as the description of their interrelations in functional structures. Another aspect which is of importance for the course of the design process is the functional flow within the product under development. The logical transition and by that the concretion (described later) of the functional model to the conceptual model is supported by the specification of the vectorial functional structure in accordance with defining the corresponding design working spaces.

1.3. CONCEPTUAL DESIGN MODELING LAYER

The conceptual design serves for the description of the solution concept of a design task. It covers all information fixed while describing the product's physical solution principles. These information contain the physical effects used for solving the problem in correspondence with the mathematical equations describing them. Geometrical information as e. g. effective lines, effective faces and effective spaces are also modeled within the conceptual design. This phase is completed by assigning the concepts modeled to the functions of the functional structure and by grouping the concepts into the conceptual structure.

1.4. SHAPE MODELLING LAYER

The shape modeling is the most concrete of the product modeling layers. For that it completes the product modeling process by the geometrical definition of all design features and design working spaces to completely defined three dimensional parts with assigned material and their combination to groups and group structures.

2. The Modeling Space of Design

By representing the design process as a process of concretion performed on different levels of abstraction (which we called in the paragraph above "the modeling layers") it may be misunderstood as a strongly forward oriented process. This is only the case in an idealized manner. In practice design evolves as a highly iterative process where, based on a solution state SS^{o}_{i}, different solution directions can be followed in order to reach a following solution state SS^{o}_{i+1}. In the following we understand by the term *solution state of an object* SS^{o}_{i} the instantiation of an object demarcated by the corresponding design working space belonging to the intended product after the i-th design step. Figure 1 gives an idea of this procedure. Starting at any solution state of a design task the solution directions showed can be followed. These solution directions which are derived by Birkhofer (1980), Krumhauer (1975) and Rude (1991) from the modeling space of design describe a possible way to transform a solution state SS^{o}_{i} into a following SS^{o}_{i+1}. In general this means to be one step closer to the intended solution.

Before stepping into more detail of the fundamentals of the design process we direct some interest on the design shown in Figure 2. It shows the presentation of a robot gripper whose design serves as example for explaining our abstract model of the design process. The robot gripper serves for the handling of small parts. It is designed for durability and for low maintenance costs. A standard connection to the robot arm was given as well as the space in which the gripper has to fit. The working method of the gripper is as follows: The force with which the handled part is gripped originates from the application of a pressure that is foreseen by the robot. The resulting force then is transmitted through a bar to a wedge at the end of the bar. The wedge splits the force into two resulting forces which are applied to the jaws of the gripper. The applied pressure causes a movement of the piston towards the jaws. The wedge at the end of the piston then causes a turning motion of the jaws which results in the gripping force of the robot gripper.

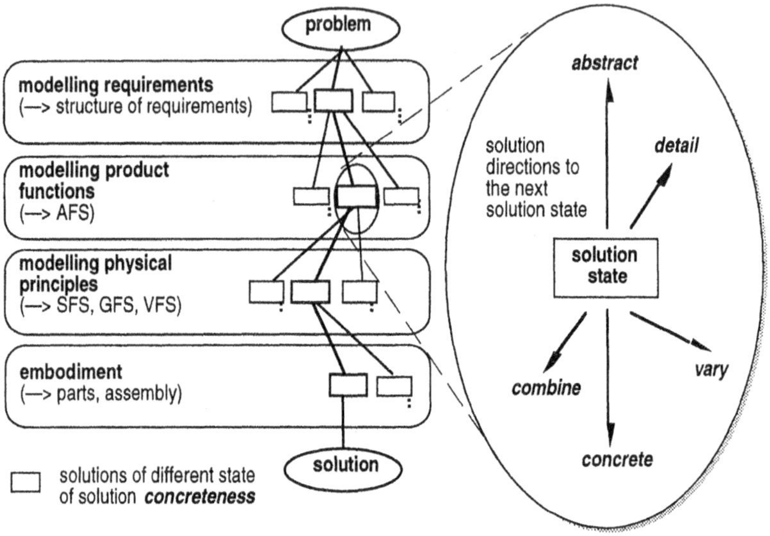

Figure 1. Elementary solution steps in the design process.

The detachment of the handled product is started by reducing the pressure applied on the piston. So a spring which is mounted in the front of the piston pushes it back to its original position. So the gripper's jaws can also be opened by a spring.

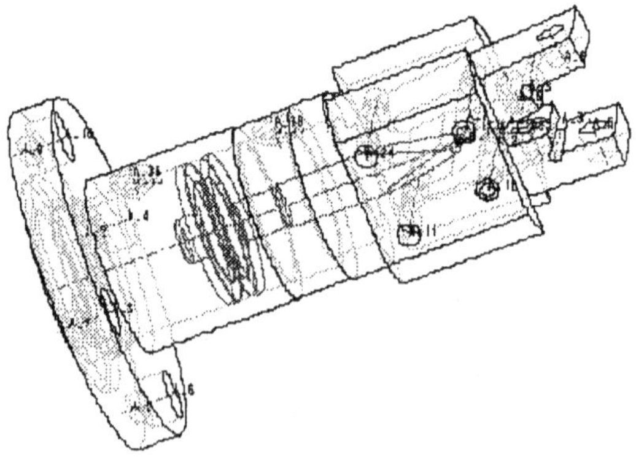

Figure 2. Product example: Robot gripper, which serves as the intended design.

In the following, the solution directions mentioned above are explained with the help of some examples from the design of a robot gripper (for a short description of the symbols used in the example please see section 3).

2.1. CONCRETION

The transformation of a solution state SS^o_i into a following, more concrete one SS^o_{i+1} is called "concretion". Here we understand by the term "more concrete solution state" the instantiation of the product model with information belonging to a more concrete modeling layer (see Figure 3). By this, new solution properties are added to the solution state SS_i. The example shows the concretion of the structure of physical principles of a robot gripper on the functional modeling layer to a conceptual design sketch on the conceptual design level.

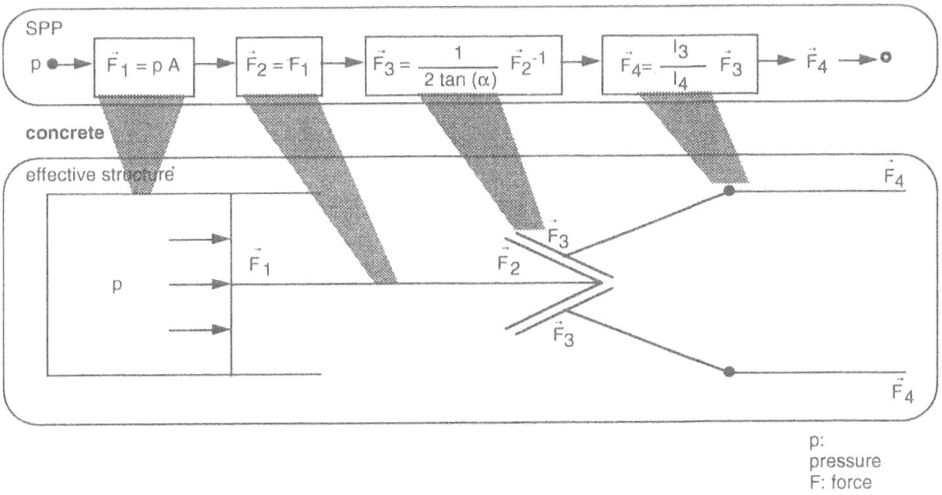

Figure 3. Example for the solution direction "concretion".

In our example the structure of physical principles (SPP) is concreted onto the effective structure of the intended robot gripper. The physical principle which describes the transformation of a pressure into a force (F_1 = pA) is mapped to the effective structure of a pressure cylinder with a piston inside. The transmittance of the force ($F_2 = F_1$) is realized by a bar which is attached to the piston. A wedge at the end of the piston splits the force into two effective forces and changes their directions ($F_3 = (1/2 \tan(\alpha)) F_2^{-1}$). F_4, which is the force responsible for the gripping action, is obtained by leading F_3 into a lever. With that a rough sketch of the effective structure is obtained. This sketch is the basis for the next design steps.

2.2. ABSTRACTION

The solution direction "abstraction" is opposite to the direction concretion. This means the solution state SS^o_i is transformed with respect to the intended design solution into a more abstract solution state SS^o_{i+1}. Abstraction serves for the recognition of essential product properties. In consequence the more abstract solution state SS^o_{i+1} is one step further away from the intended design solution (see Figure 4). This step can be used for starting from known designs to reach new until then not known solutions. In this context Figure 4 shows the abstraction of the shape of a robot gripper's jaw onto its effective structure.

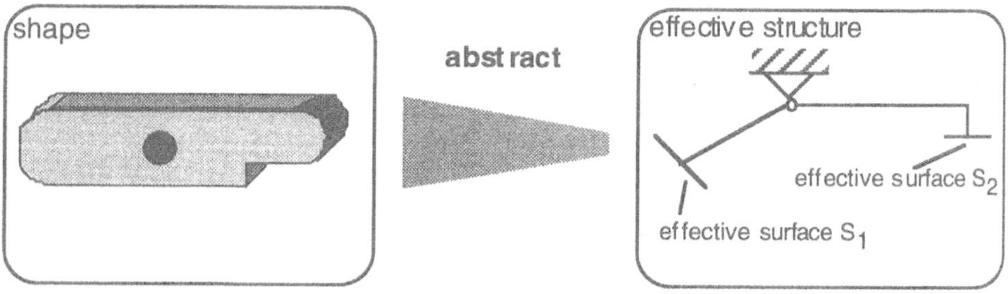

Figure 4. Example for the solution direction "abstraction".

Starting from the shape of the robot gripper's jaw, the essential features of the design are extracted. These are the two effective surfaces S_1 and S_2 where S_1 is responsible for the transmittance of a force onto the lever, which can also be extracted from the shape. S_2 is the effective surface which can be recognized as being responsible for applying the gripping force onto the handled part. The function of the hole in the middle of the jaw's body acts as a bearing which also finds its counterpart in the effective structure. By that the shape description of the robot gripper's jaw is abstracted to its effective structure. The effective structure can serve as the basis for a variation of the shape or it can be abstracted by itself in order to obtain another effective structure with other design properties.

2.3. DETAILING

Adding more information to a design object within the same modeling layer is called detailing. When detailing a solution state SS^o_i to a following SS^o_{i+1}, the modeled design information remains on the same level of abstraction as it was in state SS^o_i. The solution direction detailing is used in order to solve a design problem by dividing it into sub-problems. Figure 5 shows an example for the solution direction detailing. The rough effective structure sketch is detailed by adding two bearings to the two levers and to the bar with the function "to channel force", by adding a sealing to the piston and by designing the effective surfaces which are responsible for applying the gripping force to the part to be handled. So the effective structure of the robot gripper now contains more information than in the solution state before but the information contained in the model is still remaining on the same level of abstraction (effective structure). With all the information contained in the detailed sketch it is easier to concrete (maybe after performing other detailing steps) the effective structure to the robot gripper's shape model.

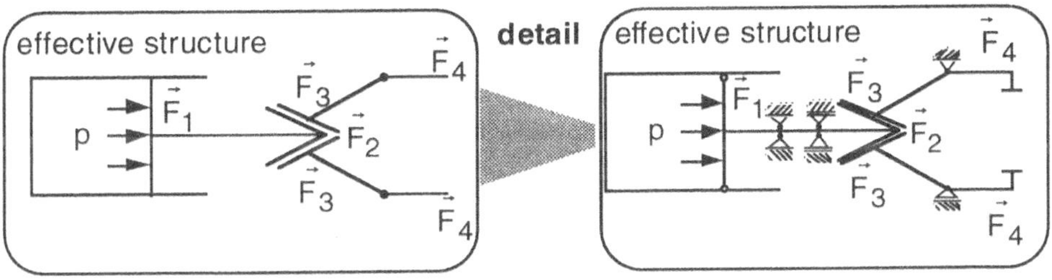

Figure 5. Example for the solution direction "detailing".

2.4. COMBINATION

This direction in the model space of design is opposite to detailing. It transforms a solution state SS^{O}_{i} into a more general one SS^{O}_{i+1}. SS^{O}_{i+1} in this case describes a summary of solution properties of the preceding solution state. One example for combination is to summarize different sub-functions into an overall function or, as shown in Figure 6, to omit different design objects in a conceptual design sketch in order to find the basis for a better variant of the intended product. Combination is also the basis for the "abstraction" direction in the modeling space of design. A design on a certain level of abstraction is combined until only the information which is essential for the next design step (abstraction or variation) is left. In our example the combined effective structure of the robot gripper's body may serve as a basis for abstracting it to its functional structure (e. g. structure of physical principles).

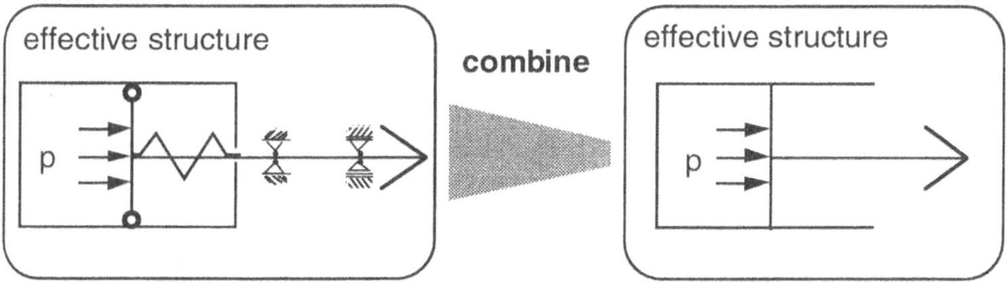

Figure 6. Example for the solution direction "combination".

2.5. VARIATION

Variation means to find to a corresponding solution, in state SS^o_i of the intended design, other eventually better alternatives (solution state SS^o_{i+1}). Variation steps back to the preceding solution state SS^o_{i-1} and concretes this solution state to SS^o_{i+1}. SS^o_{i+1} contains design properties different from the one's of SS^o_i. Variation keeps the intended design's degree of concretion unchanged. What changes, is the spectrum of possible solutions on the same level of abstraction where the design is in state SS^o_i. As an example for the here described solution direction, the variation of a physical effect corresponding to a certain sub-function, can be mentioned. Figure 7 shows the variation of the function "to generate an energy" into two variants. One is to obtain a force by applying a pressure p ($F_1 = f^1(p)$), the other variant is to obtain the force by inserting a spring ($F_1' = f^{1'}(D)$), where D is the elasticity constant of the spring.

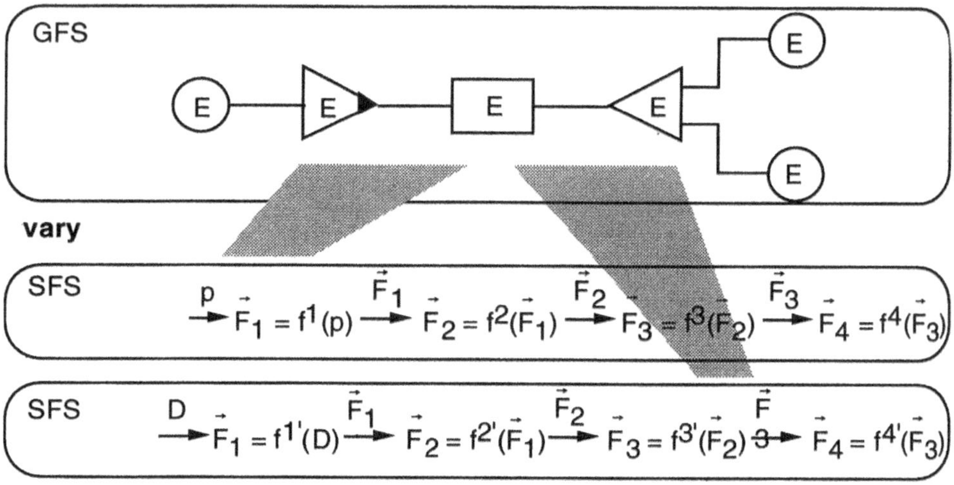

Figure 7. Example for the solution direction "variation".

2.6. GENERAL PROBLEM SOLVING CYCLE

Above we discussed the basic and elementary solution steps in the design process. Proceeding in the discussed directions after each step a problem-solving cycle has to be performed. This approach has been derived by the design methodology from the psychology of problem solving.

Figure 8 shows the basic scheme of this general problem-solving cycle (Rutz, 1985). First the designer (for reasons of better readability in the following we call the person who performs the design process "the designer" even if the term "designer/designeress" would be the more correct one) is confronted with the problem. Afterwards the definition of the essential problems is performed by fixing the objectives, main constraints and the environment for the intended solution. The next step is finding and representing a solution for the defined problem (this is the creative part of design). After that the solution has to be evaluated followed by making a decision. For one found alternative on the basis of the found solution's evaluation result. Finally as the following step of the design process, the problem-solving cycle is reiterated. So the established solution serves as a definition for the next problem. In this way, we proceed from the qualitative to the quantitative, from the abstraction to the concretion, from the incomplete to the complete etc. This general problem solving cycle is together with the design methodology, which was described above, the basis for the development of our process model of design.

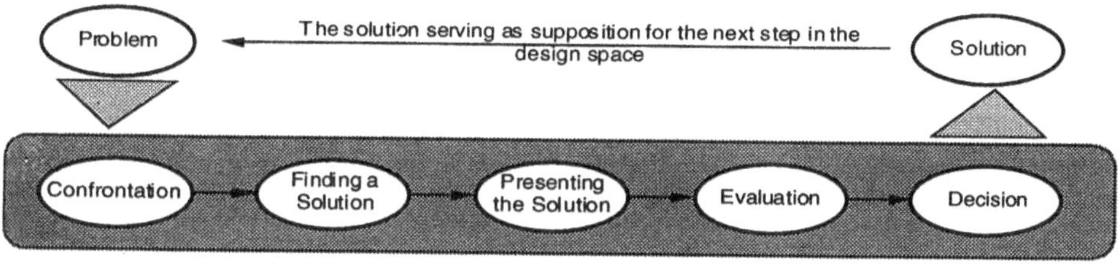

Figure 8. General problem-solving cycle.

3. Fundamentals of the Design Process for Technical Systems

With the three fundamental magnitudes of design, matter, energy and information every technical system or artifact can be described on an abstract physical level (Roth, 1994, Pahl and Beitz, 1994). Matter, energy and information are basic concepts (Weizsäcker, 1971). Information determines what has to be done to fulfill a certain purpose. Only with energy technical systems are able to perform any change in nature or in itself. Matter is the stuff, a technical system consists of. It is also the medium in which every process takes place. The human himself is the best example for a technical system. In the first instance he used tools to integrate it in his own technical system, later on he created tools themselves (that means he controlled the matter), in the last centuries he learned to control the energy and in our century, he ruled over the information processes (cybernetics). With respect to being that fundamental every design theory has to be based on this three categorical magnitudes, matter, energy and information.

Technical artifacts are connected to the environment by means of inputs and outputs and can be treated like a system. A system can be divided into sub-systems. What belongs to a particular sub-system is determined by the system boundary. With this approach it is possible to describe every technical system at every stage of abstraction. Describing a proposed technical artifact by means of a system consisting of elements, which are grouped by the system boundary related with each other by input and output, we use the term "function" or "product function". If the product function is described on the basis of matter, energy and information as inputs and outputs then we use the term "general function" (GF). If the inputs and outputs represent physical magnitudes like force or torque and the relationship between input and output is described by a physical law, then we use the term "special function" (SF). In the case of a GF the relationship between the input and output is expressed by a limited number of so called function verbs. The function verbs describe the proposed transformation between the input and output. With reference to Roth (1994) we use the set of function verbs "Change, Connect, Channel and Store" for the GF. All technical artifacts are complex constructions, so every artifact can be described by a "general function structure and a special function structure" (GFS, SFS). Because of introducing the above function types and their particular structure the fundamental working principle of abstraction is applicable and therefore a top down approach to the design process is possible.

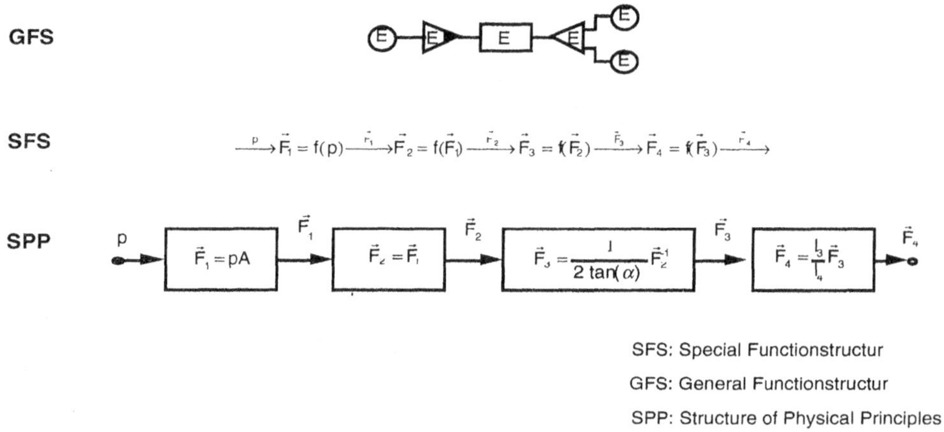

Figure 9. Established general function structure, its derived special function structure and structure of physical principles.

Figure 9 shows the established General Function Structure (GFS) of the robot gripper of Figure 2, its derived Special Function Structure (SFS) and Structure of Physical Principles (SPP). At the source of the force flow within the intended product there is stored energy depicted by the symbol of a circle containing the character "E". This energy will be changed into another form of energy. On the SFS level there is shown that the energy type of pressure will be changed in the energy type of force. After that the energy will be channeled, distributed and amplified. At the bottom of the figure the physical principles which perform this process are shown.

The knowledge about the general functions and special functions and the interrelationship between the different levels of abstraction (or design stages) is modeled in a conceptual *object model* (Figure 10). Figure 10 is intended to give an idea about the relationships between the objects of the conceptual model In this object model the design knowledge is instantiated. This means the model contains all information which the designer described for the intended product. For that reason the model contains information belonging to all design stages. If all the information described above is contained in the product model on which a design system is based, any technical product can be modeled in this system. But in the product model there is still a lack of information belonging to a mechanism which helps to structure the designed product in order to reduce the complexity of the subtasks. With respect to this problem and also focusing on the administration problem of the subtasks and the corresponding parts of the design we developed the tool of "design working spaces" which is described in the next paragraph.

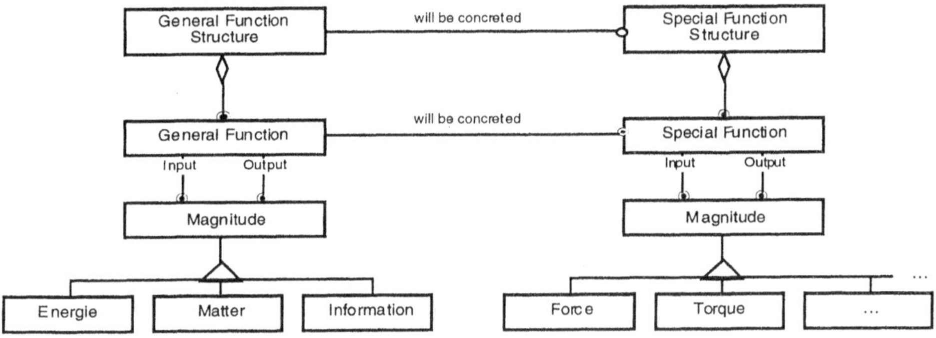

Figure 10. Conceptual object model of general functions, special functions and structures.

4. Modelling in Design Working Spaces

A design working space is a Euclidean space (on geometric level, Figure 11) which is available for the designer to solve his design task. The design working space is defined by an envelope (geometric system boundary) and its constraints (inputs/outputs). The fundamental idea of modeling design working spaces comes from system theory and therefore design working spaces are not limited to the geometry. The main purpose of design working spaces in this context of modeling design processes is to fix a special design state. If a special design state has been fixed it is possible to derive new design states getting stepwise to the intended solution.

Design working spaces are defined and will be built up by the following rules:

- A design working space consists of a set of elements and of a set of relationships between the elements.
- Elements are a set of information of the design stages, like requirements, product functions or physical principles. Relationships between the elements are general or special magnitudes like energy, information, matter or force, torque etc.
- Every design working space can be subdivided in sub-design working spaces.
- Every element, every design working space and every overall system are inside a given system boundary.
- A system boundary has one or more inputs/outputs and a function which describes the system.
- If a design working space has no inputs/outputs then we talk about a closes design working space on the other hand about an open design working space.

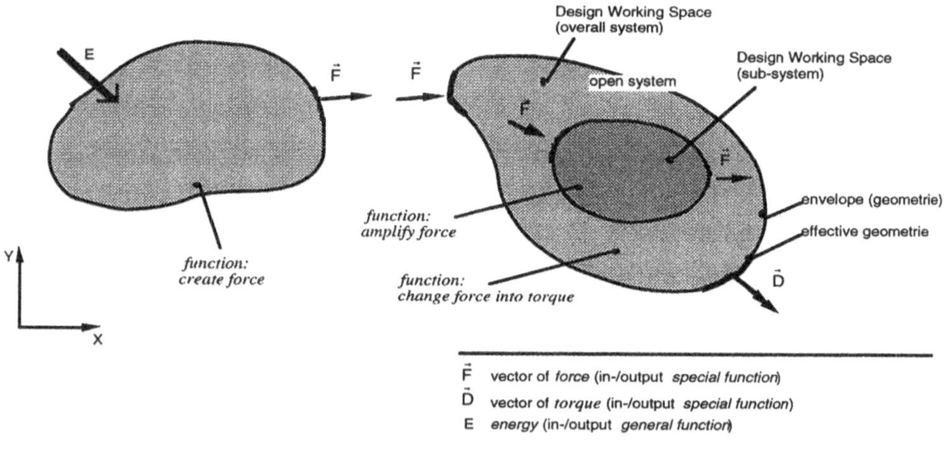

Figure 11. Basic idea of design working spaces.

In Figure 11 there are three design working spaces which have to fulfill a special product function, like create force or amplify force. The design working spaces are clearly defined by their maximum envelope and effective geometry; the envelope and effective geometry is represented by free form surfaces. The envelope describes the maximum space inside which a special problem must be solved. The effective geometry is described by effective spaces and effective surfaces which transmit for example forces. The relationship between the design working spaces in Figure 11 is established exemplarily by the general magnitude energy and the special magnitudes force and torque.

As mentioned above in this context the main purpose of design working spaces is to fix a special design state so that new design states can be derived to reach stepwise the intended solution. To derive new design states it is necessary to represent this knowledge in an appropriate model. For this the dynamic model of the design process which is described in the next paragraph, is appropriate. After that in the paragraph of a system architecture the process control model is introduced which represents special state transitions.

5. An Architecture of a Knowledge Based CAD System Based on the Dynamic Design Process

The architecture of the knowledge based design system DIICAD[1] provides four basic components. The purpose of this architecture is oriented towards the mapping of the design methodology described by Pahl/Beitz, Roth, Koller, Hubka and others onto a CAD system.

The first component is the "object model component". It is responsible for the description of the design task as well as for its solutions. All the information describing the intended product is contained in DIICAD's object model (see also paragraph 3). First the requirements, which the product has to fulfil are modeled. This so called "requirement model" then serves as the basis for the following design process (Kläger, 1993). The most abstract level of design is the functional modeling. Herein the functions, a product has to perform, are modeled in a very abstract way. The design methodology provides as the next level of modeling the conceptual design. In conceptual design there are two different views on the product. The first assigns the physical effects to the respective functions of the functional level. The other maps the functions onto the principle structures of the product. The product's effective structure is the result of the conceptual design phase. The functional model of a product in connection with the physical effects and the principle structures form together with the requirements model the basis for the shape design. So all design information needed is stored in the object model.

The second component of the system architecture is the "task solution component". Within this component there are processes defined which transform one solution state SS^{o}_{i} of the design into a following solution state

1 DIICAD: Dialog oriented Integrated Intelligent CAD System

SSo_{i+1}. Currently there is a case based approach to the task solution component. A learning and self-controlling approach has not yet been attacked. Until now this learning and hopefully self-controlling "component" is the designer himself.

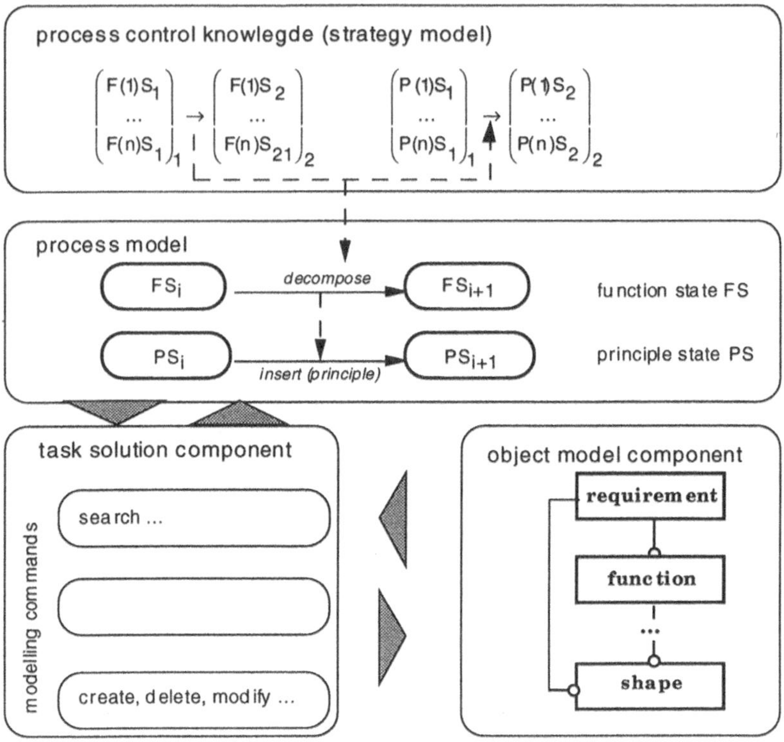

Figure 12. Architecture of the knowledge based CAD systems.

The third component of our system architecture is the "process model component". It contains a dynamic model of the design process. This dynamic model contains the objects of the object model. It contains also the different states in which the objects can find themselves. The dynamic aspect now describes the possible transitions between the different states of the objects and, as a very important point, by which actions these transitions are caused. The dynamic model is used by the process model component in order to compute the path of design steps leading to a description of the intended product.

Finally the fourth component of the system is the "process control component" which is responsible for the design path leading from the requirements specification of an intended product to its shape modeling.

The process control component evaluates the status of the current design and computes on the basis of the process model the next design step which should be performed in order to transform a solution state SS^{O}_{i} into a following state SS^{O}_{i+1}. The information computed here contains on the one side the design object which should be manipulated in the next step, on the other side it contains the solution direction in which the design process should step forward. The process control component contains different general strategies which lead through the design process. The most important strategies which are modeled is on the one side the course through the design along the main functions of the intended product and on the other side the strategy "design along the functional flow" can be applied. In each case the process control component causes the modeling of the requirements of the intended product. After finishing the requirements specification, the functional modeling is initiated. When the functional structure of the system has been modeled, the designer chooses one strategy which seems to him to be an appropriate way for coursing through the design process. By the help of the chosen strategy, which is modeled in the system and together with the process model, the system calculates the design object which should be manipulated in the next step. Also the solution direction (see section 2) to be applied is chosen. So the process control component passes the reference to the design object and the solution direction to the task solution component which is responsible for manipulating the instantiated object model of the design.

It is understood that the general strategies currently contained in the process control component are not sufficient for completely controlling the design process. For that reason the research work is directed towards a learning design process controlling system. This additional approach takes the task dependent knowledge into account. This means the system has to learn the design steps for a specific task from the designer. Those steps have to be stored in combination with the respective solution states of the intended design in order to retrieve them when a design with similar requirements and functional structure has to be performed.

5.1. THE DESIGN PROCESS MODEL

Depending on the experience and skill, a designer chooses the appropriate path for the particular steps at different design stages. One important point for the development of an intelligent design system is to develop a model which describes this design process knowledge in a computable form. This means that an intelligent design system must support the designer in finding the right way through the design by navigating through the modeling space of design. For that reason the possible solution directions, applied to the objects of the respective modeling layers, described in section 2, have to be modeled in a so called dynamic model of the design process. This model shows the dynamic behaviour of the design objects over time.

5.2. THE DYNAMIC DESIGN OBJECT MODEL

The design process model consists of two components. One is the dynamic model which describes exactly, thus in a computable form, the dynamic behaviour of the design objects during the design process. This means the states which an object can assume are modeled as well as the state transitions of the object. The second is the process control model, described after the dynamic model. The contents of a dynamic model are shown in Figure 13 and are described in the following.

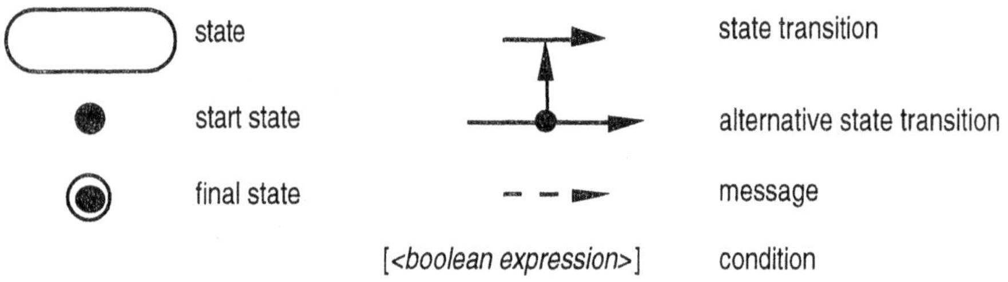

Figure 13. Symbols used in dynamic modeling.

Figure 13 shows the different symbols used in a dynamic model. The symbol "state" stands for the state which the modeled object can assume. An instantiation of the state, for example, can be detailed or varied. In the dynamic model this means, the object is in the state detailed or varied, respectively.

There are two other special states in the modeling method. These are the start state and the final state. The start state characterizes the object in its starting point before its first instantiation, whereas the object is in its final state when all attributes are defined and consolidated so that in this design step no further work will be done again on this object.

The arrows connecting two states symbolize the state transitions. A state transition is always caused by an action which designates this transition.

A special kind of state transition is the alternative state transition. This means that if the main action which designates the respective transition cannot be performed, the alternative transition caused by the corresponding alternative action is performed.

The last type of items which are used in dynamic modeling is the message. When an object reaches a state, which was specified in the model, a message can be sent in order to start an action. For that reason the dashed arrow which is the symbol for the message type always points from a state type to a state transition type.

In the case when two transitions which start from the same state are possible, the specification of conditions (symbolized by a rule enclosed in brackets) is necessary. So with the help of conditions it is possible to specify the transition which is only followed in the case in which the respective condition is fulfilled.

Figure 14 shows the dynamic model of the object general function. As the general function specification layer after the design methodology described in paragraph 2, is the lowest layer in the concretion hierarchy, the design starts with the description of the general function of the intended product. When the requirements which the intended product has to fulfill are specified, the object General Function (GF) is in its initial state. Here the action initiate is performed. By initiation the object GF changes its state from the start state to the state defined. The two transitions starting at the state defined, symbolize that there are two possibilities for the next transitions. One is the transition to the state varied, the other is the transition to the state defined. The transition to the state varied is marked with the condition [GF.detailed ≠ NULL]. This means, this transition is only then followed if the instantiation of the object General Function has already been detailed before.

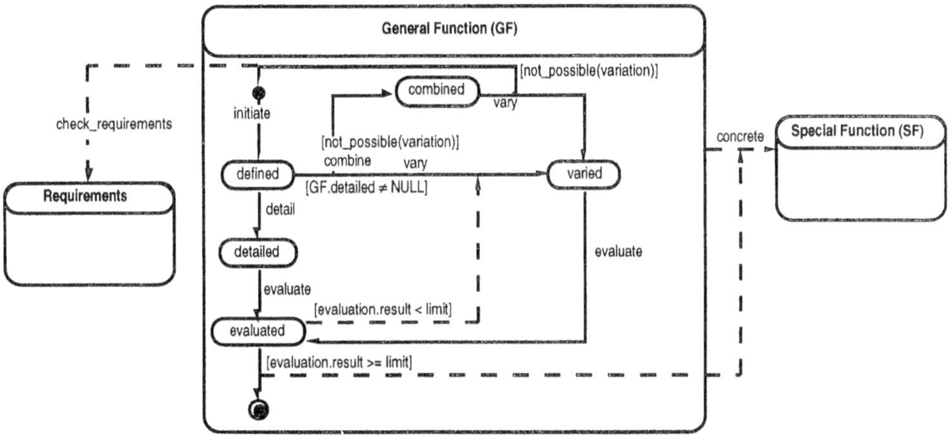

Figure 14. Dynamic model of the object "General Function".

In consequence, if General Function has not yet been detailed before, the transition to the state "detailed" is accomplished. The new state, detailed, is then reached by the action "detail". After detailing, the results of the action must be evaluated (how the evaluation is performed will be described in the following). By this the state "evaluated" is reached. The evaluation calculates a result (evaluation.result) which describes the quality of the found solution which was gained by the design step currently performed. The result of the evaluation is then compared with a certain limit. If the result is better than the limit, the final state the object general function is reached. This means the design for the general function is accepted. Here a message is sent in order to cause the sending of the message "concrete" to the object Special Function (SF). This means, the general function object is concreted to the corresponding special function.

If the result is worse than the limit, a message is sent to the transition *vary*. This message causes the execution of the variation of the general function object. In our model, the transition "vary" is modeled as an alternative transition. This means, if the intended variation can be performed, the state "varied" will be reached. Then the properties of the object will be evaluated in analogy to the object in the state detailed. If the variation cannot be executed (symbolized by the condition [not_possible(variation]), the alternative transition is started. This alternative causes the combination of the object so that the object will reach the state "combined".

When the object is in the state "combined", the transition vary is performed. In the case when a variation is not possible, no useable result has been reached in this design step. In consequence, the object general function returns into its start state, while the corresponding transition sends the message "check_ requirements" to the object requirements.

Similar to the dynamic behaviour of the general function object is the behaviour of the other objects which belong to the different modeling layers (see also Figure 15).

Figure 15 shows on overview of the dynamic behaviour of the objects which classify the different modeling layers in design. The internal dynamic behaviour of the objects shown (special function, physical principle, effective structure and embodiment) is very similar to the general function. The difference is that if no solution for the specified problem is found, the objects are abstracted to each next higher level of abstraction. This abstraction can be executed until the highest level, the general function modeling, is reached. Here an abstraction is no more possible. If no solution is found, as shown in Figure 14 the requirements have to be checked for correctness. If the intended design solution has to fulfill all the specified requirements, and no appropriate solution can be found, in consequence, the design project has to be stopped.

The process model represents all states, state transitions etc. which are described in the modeling space of design being fundamental for navigating through the design process. A special "path" for particular steps is not described in the dynamic design model but in the process control model in which knowledge of designers, their experience and skill is described. This process control model controls special state transitions and messages of the dynamic design model.

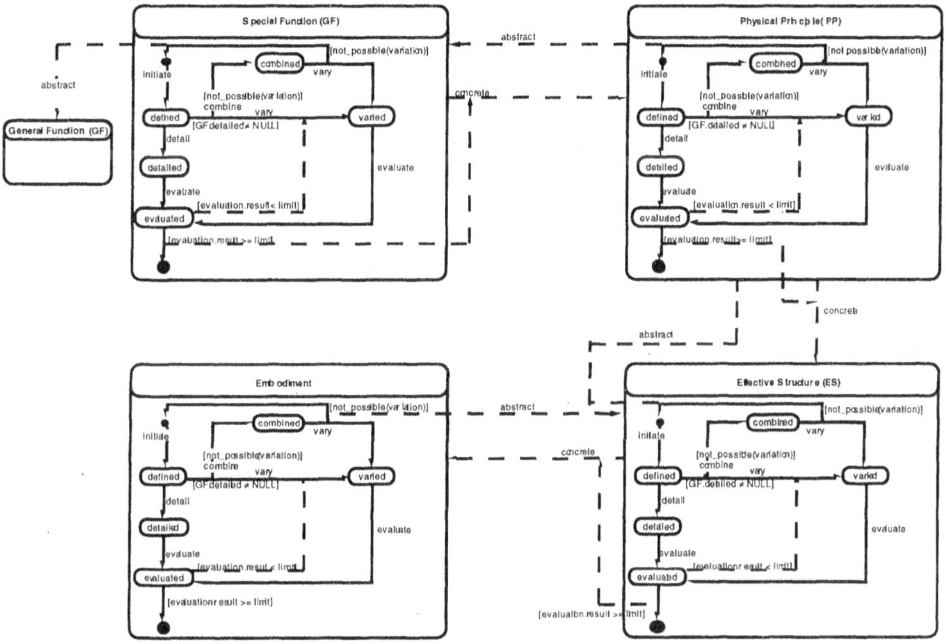

Figure 15: Overview of the dynamic model of the objects which describe the abstract modeling layers in design.

5.3 THE PROCESS CONTROL MODEL

The dynamic design model, described above, is responsible for the selection of the solution directions i. e. the modeling commands applied next to a given design object in a given design working space. This procedure can be called a "tactic" which is subordinated to a superior design strategy embodied by the process control model. So the process control model contains several design strategies which describe the design process on a higher level than the dynamic design model. These process control strategies are heuristics which describe a path leading to the solution of a design task. The way resulting by the strategies usually is not optimal and by that it does not lead directly to the intended design solution. Partially iteration cycles are necessary in order to improve already found solutions or to correct directions followed by mistake. Because of their taskindependencies, process control strategies form the basis for a process control of an intelligent CAD system. This means they describe independently from the specific design task a path through the design leading to the intended task solution. The knowledge which is necessary for this task is based on the general design methodology. As example for such strategies "design along the flow of force, begin at the source" or "design along the main functions" can be mentioned. It is easy to see that these general, task independent strategies can show any path leading to the solution of the design task. But this path usually will not be optimal. In this context the term optimal means to design a product which fits optimally to the requirements specified applying possibly few design steps. Finding such an optimal path needs knowledge about optimal design processes depending on the respective design task. For that reason in our research work we try to develop a concept to acquire task dependent process knowledge from the designer in order to reuse it when a similar task appears.

6. Conclusions and Future Work

We have modeled and verified the dynamic model in a small application on the DIICAD product model. Modeling design solutions in design working spaces and saving these design solutions as solution patterns in the product model is possible. With design working spaces we find similar solution patterns for a given problem. This is realized by using a case-based-reasoning approach which is implemented on KEE and ACIS. At the moment we are able to do this for the requirements modeling (Kläger, 1993) and the functional (Huber, 1994) design stages in a top down approach and for solution patterns (Suhm, 1993) described in the mentioned PhD theses of Kläger (1993), Suhm (1993) and Huber (1994).

Modeling the design process in the described way is a promising approach. We have modeled the product life cycle and verified the approach in a small prototype. In the SFB346 (this is a special research area which is set up at the University of Karlsruhe by the German Research Community) a language has been developed to describe dynamic models. We consider it an important point that in the future basic research has to be done in developing a methodology to build dynamic models. Our next step will be to implement the process model as a whole (dynamic design model and process control model) which will be controlled by the process control model. Another step is to develop a methodology to gain design process knowledge from the designer by configuring so called process patterns which are analogously the same on the dynamic level as solution patterns (Suhm, 1993) on the static level.

References

Benz, T.: 1990, *Funktionsmodellierung als Basis zur Lösungsfindung in CAD-Systemen (Functional Modeling as a Basis for the Solution Finding in CAD Systems)* Universität Karlsruhe (TH).

Birkhofer, H.: 1980, *Analyse und Synthese der Funktion technischer Produkte* (Analysis and Synthesis of the Function of Technical Products), Dissertation, TU Braunschweig.

Huber, R.: 1994, *Wissensbasierte Funktionsmodellierung als Grundlage zur Gestaltsfindung in Konstruktionssystemen /(Knowledge based functional Modeling as Basis for the Shape Computing in Design Systems)*, Aachen: Shaker (Reihe Konstruktionstechnik), Zugl.: Karlsruhe, Univ., Diss., Institut für Rechneranwendung in Planung und Konstruktion (RPK).

Kläger, R.: 1993, *Modellierung von Produktanforderungen als Basis für Problemlösungsprozesse in intelligenten Konstruktionssystemen (Modeling of Product Requirements as Basis for Problem Solving Processes in intelligent Design Systems)* Aachen: Shaker, (Reihe Konstruktiostechnik), Zugl.: Karlsruhe, Univ., Diss., Institut für Rechneranwendung in Planung und Konstruktion (RPK).

Koller, R.: 1985, *Konstruktionslehre für den Maschinenbau* (Design Theory for the Mechanical Engineering), Springer-Verlag, Berlin.

Krumhauer, P.: 1974, *Rechnerunterstützung für die Konzeptphase der Konstruktion* (Computer Support for the Conceptual Phase of Design), Dissertation, TU Berlin.

Pahl, G. and Beitz, W.: 1994, *Engineering Design*, Springer-Verlag, Berlin.

Roth, K.-H.: 1994, *Konstruieren mit Konstruktionskatatogen*, Springer-Verlag Berlin.

Rutz, A.: 1985, *Konstruieren als gesdanklicher Prozeß (Design as Intellectual Process)* KM Lehrstuhl für Konstruktion im Maschinenbau TU München, Dissertation.

Suhm, A.: 1993, *Produktmodellierung in wissensbasierten Konstruktionssystemen auf der Basis von Lösungsmustern (Product Modelling in Knowledge Based Design Systems on the Basis of Solution Patterns)* Aachen: Shaker, (Reihe Konstruktionstechnik), Zugl.: Karlsruhe, Univ., Diss., Institut für Rechneranwendung in Planung und Konstruktion (RPK).

von Weizäcker, C. F.: 1971, *Die Einheit der Natur-Studien*. (The Unity of the Nature - Studies), Hanser, Munich.

Wallace, K., Ball, B. and Tang, M-X.: 1995, AI in mechanical engineering design, *in* J. S. Gero and F. Sudweeks (eds.), *Fourth Workshop on Research Directions for Artificial Intelligence in Design*, University of Twente, Enschede, The Netherlands.

13

FORMAL SUPPORT METHODS IN DESIGN: DISCUSSION

STEPHAN RUDOLPH
Stuttgart University, Germany

1. General Considerations

Computers have affected many areas of our lives since their first appearance. The changes induced by computers have not only affected our way of doing things or monitoring and controlling processes. Computers have also changed our perception of the world and the way we think. For this reason the existence of computers as modern production means has had a significant influence on our way of doing science and design research as well. The modern paradigm "design is search" versus the paradigm of the former century "design is composition" accounts for this change of views and methods of many engineering design researchers. Computers have become the medium of choice of many.

However, despite the widespread applications of computers in virtually every area of engineering and design, little importance has been attached to fundamental, not so obvious but nevertheless important, implications of computer usage. A closer look at these fundamental implications of computer usage in general and especially in the area of engineering design reveals the following three main issues.

1. *Representation.* In order to write a useful program, a set of operands and operators needs to be specified. They are intended to be representations of real objects and the permissible manipulations on these. As with any representation, the question occurs as to what extent this representation is complete, and, since all operations occur on these operands and are thus limited to these, whether the underlying *closed-world assumption* is valid.
2. *Mapping.* Computer programs generally compute a certain set of output data O from a certain set of input data I. Mathematically speaking, a computer program represents an algorithm P which maps the input data onto the output data, thus $P : I \longrightarrow O$. However, the set of all algorithms P is a true subset of all mathematically imaginable mappings M, thus $P \subset M$. Computers can therefore in general only deal with problems which are *algorithmically solv-*

able. This is a major restriction on the kinds of problems that computers are capable of dealing with.

3. *Reasoning*. Formal models of logic rely heavily on certain characteristic properties such as proofs of consistency, completeness or correctness. However, any extension of theoretical models is only feasible when through comparison of the natural world with the model predictions discrepancies are identified by observations or experiments. Because of the continuing lack of devices serving as computer interfaces suited for such needs, these inconsistencies cannot be detected and conceptualized automatically. This seems, at least for the near future, to remain a domain for the problem identification and problem solving capabilities of humans.

The first and second of these three main issues are today more or less well identified and understood. They have been encountered and tackled since the appearance of the first implementations of numerical algorithms in computer programs. Take, for example, the consequences of the *finiteness* of the representation of numerical values of program objects, i.e. of the program variables, are all kinds of stability and convergence problems in numerical algorithms and originated much research devoted especially to the solution of these problems. Today it is generally accepted that the fact of doing computation on machines with finite precision, i.e. incomplete representation, lies at the heart of many problems and the trend to symbolic computation can be interpreted as one way to try to circumvent this problem at the expense of the introduction of other difficulties.

The third major issue is of a more difficult nature and not so obvious. For this reason it is investigated here in some more detail. The continuing and correct execution of a set of two operations is studied in Figure 1. In this figure, the outcome of a surface walk of a person (the object) to which two different operations can be applied (walking infinitesimally 'straight' for a certain amount of time and then making a ninety degree left turn), is drawn. As shown in the left hand part of Figure 1, going 'straight' on the surface of some (flat) plane from a point A to a point B, making a ninety-degree turn at B, going then 'straight' from point B to a point C, making a ninety degree turn at C, going on the surface a third time 'straight'

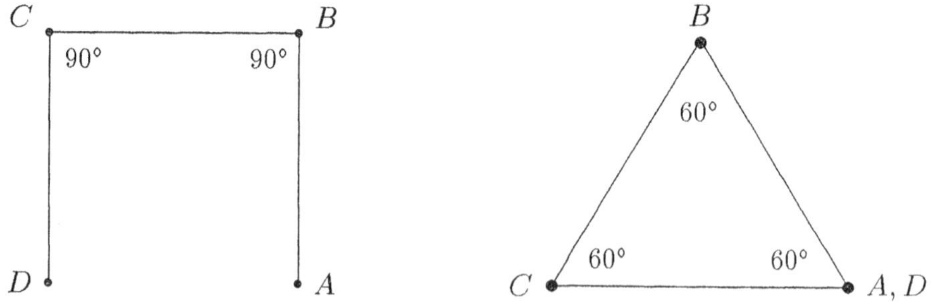

Figure 1. Two walks on a 2-dimensional flat surface

from point C leads finally to point D, which is (expected to be) quite distinct from the point A of departure. If however the point D turned out to be identical to A, one realizes that there is something wrong — this can only happen when the angles are sixty and not ninety degrees, as shown in the right hand side of Figure 1.

However, it is also imaginable that another reason lies at the origin of the difference in the expected outcome of the surface walk. This is shown in Figure 2, where the surface walk happens to take place not on a 2-dimensional plane, but on the surface of a 3-dimensional sphere. There it is easily possible to arrive at point A again, after walking 'straight' three times and making ninety degree turns at B and C respectively. Contrary to humans, computers cannot 'experience' such discrepancies in computer program simulations, since the concept of the dimensionality of the surface has to be a part of the *a priori* chosen representation defined by the programmer. Since experience is inseparably tied to reality, while simulation depends

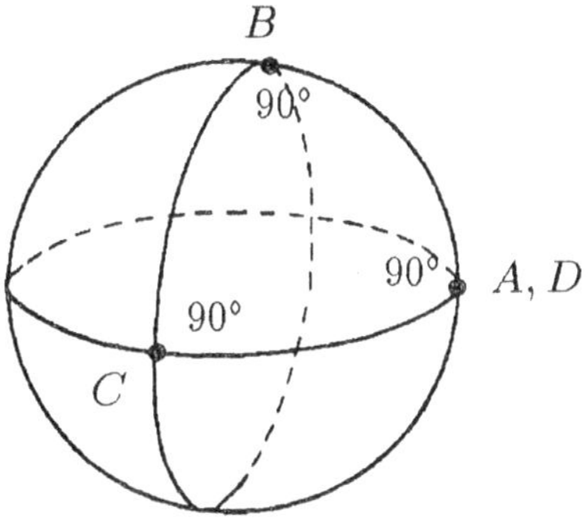

Figure 2. A walk on a 3-dimensional sphere

on a representation of the latter, it is important to realize that both the experience of 'inconsistency' as a result after executing a set of 'correct' operations, as well as an attempt to formalize the occurrence of such an inconsistency into an extension of the currently valid (geometrical) theory to a more general theory, seems very hard if not even impossible to automate. A general theory, which does explain the new experiences, has to contain and explain all former experimental knowledge as special cases as well. It is strongly questioned whether computers are generally capable of doing such truly conceptual generalizations. If computers were able to, then one could argue that one could create a program which could become monotonically 'smarter', just by executing itself. Since all the operations inside a program rely on the inbuilt representations provided *a priori* by the programmer, in reality the underlying metric of the space does change automatically and without notice, since it depends only on the individual problem area under consideration.

While these considerations might seem pointless or even superficial to many, the strong connection of these thoughts to the superior capabilities of humans over computers to form novel theories and to be innovative in an engineering sense should be evident. These are capabilities which do heavily rely on a personal experience of 'inconsistency', which can be strongly subjective and does not even animate two different humans necessarily to come to the same conclusions. Experience shows that 'inconsistency' can pass unperceived and it is possibly just this which makes true innovation such a rare event.

While keeping these principles and fairly general implications of the use of computers in mind, the next section investigates in more detail the various specific effects of the use of computers in engineering design today.

2. Computers in Engineering Design

Besides the general theoretical implications of computer usage discussed in the previous section, the existence of computers has also affected the practice of doing science and engineering design research. In former times science seemed to be in principle an iterative sequence of empirical and careful observation of natural phenomena, its modeling and then a continuing comparison of the model predictions with further observations. Today, the focus is often shifted to a methodological sequence of the definition of axioms and the specification of a set of permitted operands and operations (i.e. a formal logic based on the axioms), which is then used to derive a set of statements as formally derivable consequences of a certain sequence of operations.

While on the one hand this approach offers the advantage of being formal and thus able to be objectified and programmable, there is on the other hand also the danger of losing the vital link between the formal (design) model axioms and the reality of the real (design) world. At the worst this may even lead to clean, nice and powerful formal (design) models and methods, which have nothing to do anymore with the real (design) world.

This motivates the question how the usefulness of newly suggested design models or methods can be evaluated. Two possibilities were discussed without agreed conclusions.

1. *Benchmarking.* Arguments in favor of benchmaking included the transparency of the evaluation process through a clear statement of the performance criteria, and the need for such a procedure for comparison of newer design methodologies with older ones. Arguments against benchmarking included the impossibility of stating generally valid performance criteria. It was argued that a particular design solution may not be determined so much by the choice of a certain design procedure (i.e. the design methodology). as by the existence of problem specific boundary conditions. By their nature, these boundary conditions do largely differ from case to case. Additionally, there might be an unconscious tendency of humans to evaluate and compare the outcome only, i.e. the design solutions, and not the differences in the design process

itself as a consequence of another axiomatic foundation of the used design methodology.

2. *Problem Collections*. As an alternative to the preceding point, where it turned out that benchmarking of design methods seems difficult to carry out, the use of design cases as case studies was suggested to help the comparison of the different aspects and procedural steps of the various design methods with one another. However, it was argued that it would be very difficult, if not impossible, to establish such a required reference collection of meaningful design examples.

While the discussers felt that benchmarking and case studies are valid tests in many other areas of engineering, its application to engineering design would be advantageous to increase transparency and objectivity in the field, if the above mentioned difficulties could be overcome. As a first step in this direction, it was suggested to look at the collection of the different design methods as a *toolbox*, where each tool might be ideally suited for its specific purpose. On the other hand, the very same tool, ideally suited for a specific purpose, may perform only adequately or even badly for other purposes outside its original scope.

The previous arguments show that it is important to characterize the precise scope and applicability of any newly developed design theory or method. The various theoretical and practical limitations of the method used should also be clearly identified. The set of all critical underlying assumptions of the method should also be explicitly stated. A thorough treatment of these theoretical and practical issues is not only a question of individual scientific honesty and the personal working style of any researcher, but is essential for any true scientific advancement in the field of engineering design. It also assists the necessary characterization of the various design methods and methodologies by stressing their individual differences versus their commonalities in the above sense.

The other important point in the establishment of computational theories of design seems to be the nature and the choice of a set of appropriate axioms. Since any statement formally deduced in such a logical framework can be shown to be based on the previously chosen set of axioms, the question arises how much validity and unforseeable consequences are inherent to the choice of a certain set of axioms. Is their choice really free and arbitrary? From a purely theoretical point of view the answer is: Yes, of course! The main concern of theory building is the formal deduction of statements using some sort of established logic.

On the contrary, from an epistemological viewpoint, the answer to the very same question is: No, of course not! In thermodynamics for example, it is always observed that the heat flows from the hotter object to the colder one, and never in the reverse direction. If the empirical observation of such a unique behavior had not been made, this principle could not have been established. Thus, the second principle of thermodynamics is not arbitrarily 'defined', but is in fact completely epistemologically based and justified. It is just this epistemological basis of the axioms which is the vital link of the established theoretical model to reality and provides in large part the validity of all the subsequent formal deductions by means of this theory.

What, then, are the observations in every day design reality which could be used as an epistemological basis of design and formulated as its axiomatic foundation? This is a tough question to answer. For the moment, there seems to exist no common agreement in the engineering design community on such an axiomatic foundation in the area of engineering design. There are many existing paradigms in the form of *Design for 'X'*, i.e.

- Design for Manufacturing,
- Design for Assembly,
- Design for Disassembly,
- Design for Recycling, etc.,

which do summarize and characterize some basic observations about the design process. However, the above paradigms are by no means formal necessities in a mathematical sense. For this reason, they may never qualify for a true axiomatic foundation of engineering design from an epistemological viewpoint. It seems that this fact of a still unidentified epistemological foundation of engineering design represents one of the most important current bottlenecks of design research. Any significant contribution in this respect would be an important theoretical contributions to the development of a future science of engineering design sought by many.

If the identification of an epistemologically justified axiomatic foundation of a theory is such a difficult thing to do, what general heuristics do mathematicians and scientists use when they have developed a new theory and try to judge its usefulness? In such a case one may resort to one or more of the following criteria: Is the theory

- powerful?
- simple?
- elegant?
- beautiful?

In fact, these criteria reflect the formal embedding of logic in aesthetic and of aesthetic in ethic. This means that the existence and the outcome of the former can be explained and justified by the latter. This philosophical view refers to ancient Greek philosophy and dates back to Aristotle. A sketch of this mutual embedding of logic in aesthetics and of aesthetics in ethics is shown in Figure 3.

So why are computers, despite the drawbacks discussed, such attractive research and application tools? The answer seems to be obvious. Once a software program exists, the program and its data can easily be duplicated, distributed and shared with others. Once programmed, the high execution speed of relatively simple operations provides an alternate way to explore possibilities and simulate the behavior of objects which have never existed in reality. Search techniques, the paradigm

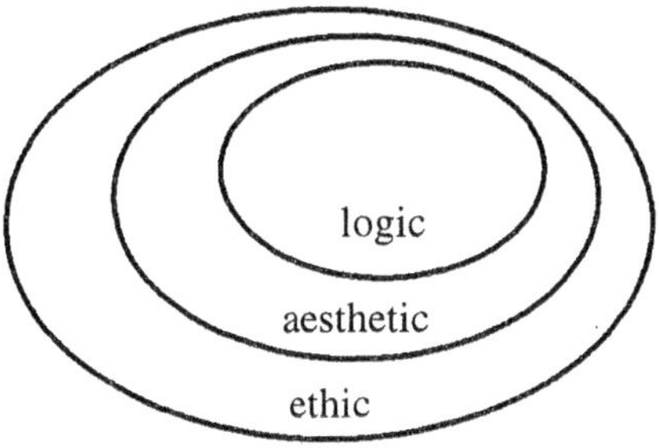

Figure 3. Aristotlean philosophy

"design is search" and simulation program packages using finite element techniques are well known and accepted examples of this.

The following two papers of this workshop session show besides their individual attempts to contribute to the advancement of the field, examples of individual compromises in terms of the above issues as well. The discussion of the two papers after their presentation during the workshop session is summarised below.

3. Paper Discussion

The paper *Formal Concept Analysis in Design* by Mihaly Lenart describes the use of Hasse diagrams for the analysis of complex situations in order to facilitate decision making in design. Based on a matrix containing the relations between the design objects and their properties, a Hasse diagram is constructed founded on the theoretical concept of lattices.

Possible benefits of such a purely formal approach are the apparent generality of the method and its independence of a particular problem area. The Hasse diagrams provide a transparency of the coupling between objects and properties. The representation of this coupling also allows the identification of a hierarchical order. The representation form can be nested to hide details and to deal with a large number of edges and vertices. In terms of computational complexity, an exponential growth is expected on average, but nesting can be used to reduce the number of displayed edges and vertices.

Possible disadvantages of the method are the missing representation of design contexts in this context free method. Also, due to the lack of familiarity of most designers with theoretical concepts such as lattices and Hasse diagrams, the two questions of the adequacy of the method concerning the requirements in engineering design and of the demonstration of the practical usefulness of the theoretical approach, could not be answered.

As a future perspective, the author was encouraged to demonstrate the suggested technique using more realistic engineering design examples and to clarify the discussion of the usefulness and limitations of his approach to design analysis and/or design synthesis.

It was interesting to see that strong formal methods can sometimes also suffer from their formality when applied. Abstract concepts, like the theoretical concept of lattices, make it sometimes more difficult to convince prospective users to make the required extra effort to become familiar with a particular mathematical technique. Due to the 'inertia' inherent in many humans, new techniques are often only adopted when they can handle tough cases and solve important problems other techniques have been shown repeatedly to be incapable of solving. This seems to be one of the major obstacles for many new methods and represents one of the more important factors slowing down the adoption of methodological change and innovation.

In contrast to the formal theoretical concept of lattices, the second paper of this session relied on the known and already well established systematic German engineering design approach developed by Beitz, Ehrlenspiel, Hubka, Koller, Pahl, Roth and others.

The paper *Supporting the Design Process by an Integrated Knowledge-Based Design System* by Hans Grabowski, Ralf-Stefan Lossack and Clemens Weis describes the embedding of the German design methodology into a framework of a knowledge-based CAD system. German design methodology considers the design process as an iterative process of concretization performed at different levels of abstraction. These different levels of abstraction are classified into the following.

1. *Requirements modeling layer*, containing the preconditions of the design, the definition of the product requirements and the description of the product's immanent task structure.
2. *Functional modeling layer*, serving to represent the functions and the functional interrelationships of the design objects to be developed.
3. *Conceptual modeling layer*, containing all the information to describe the solution concept of a design, such as physical solution principles, effective spaces and the grouping of the functional structure into a conceptual structure.
4. *Shape modeling layer*, completing the above product modeling layers by the geometrical definitions to fully define the 3-dimensional parts with assigned material properties and their combination into a part group structure.

In each of the above modeling layers, the following activities can be performed to advance the current solution to the next solution state. These are the steps to

- detail
- vary
- concretize
- combine or
- abstract

the solution state.

Starting with this description of the major elements of the German design methodology and the classification into the above modeling layers, the possible future integration and support of this phase oriented view of the design process by means of a knowledge-based CAD system was described. It was suggested this methodology be provided with a conceptual specification of how to navigate through the design process using a dynamic process model. The complete envisioned approach, including the still missing knowledge based modules for some of the four modeling layers, was demonstrated using the example of a more and more detailed design of a robot gripper. A prototype of this system is already partially implemented as a working software tool.

Possible benefits of such a systematic design method lie in the systematic decomposition of the complex design process, which helps to suppress hidden assumptions. Further, the systematic investigation and use of physical principles might help to guide the search for novel design solutions. This is a similar procedure to the morphological box approach, where physical principles are systematically combined to create new conceptual design solutions. Finally, the idea to support the designer by means of case-based knowledge to help the navigation through a dynamic design process model seems to be promising.

Possible limitations of the method might be its restriction to specific problem areas where functional modeling using physical principles is applicable. Also, the necessity to specify the definition of the input/output properties of the future design object might be difficult in some complex design cases.

4. General Outlook

Design objects are a part of our daily environment. In this respect, design objects have many consequences for humans. One could even say that design objects and thus design affects society: the invention of the automobile has transformed our society in the last one hundred years significantly. The existence of cars has created the need for roads, highways and parking lots. This affected cities as well as the shape of our countrysides. Even the organization of our cities is affected by the consequences of the ability to be readily mobile with personal cars.

Since most events are more likely to be coupled than isolated events from one another, one can say that society affects design at least as much as design affects society. The creation of design 'styles' and the existence of laws, design norms and expressions like 'political correctness' show the closure of this feed-back loop.

Additionally, the growing competition in world wide consumer markets increases the pressure on virtually every company towards new, modern and more innovative products. Besides the two aspects of innovation in engineering designs, e.g. product-driven designs and technology-driven designs, what is the true source of innovation? Is there one common concept of it? Can innovation be algorithmically formulated or enforced? These are some of the fundamental and driving questions of engineering design research today which make it such a challenging field with possible benefits to many other scientific disciplines. Determining the theoretical issues of the possibilities and the impossibilities of computer support in engineering design has therefore been one of the many interesting topics of this workshop session. However, remarkably many final answers to this problem of humankind still have to be searched for.

PART FIVE

Design Process Methods

14

A LOGICAL THEORY OF DESIGN

FRANCES BRAZIER, PIETER VAN LANGEN AND JAN TREUR
Vrije Universiteit Amsterdam, The Netherlands

Abstract. Design tasks typically deal with incomplete information and involve flexible reasoning patterns for which sophisticated control strategies are needed. As a result, the reasoning patterns are highly dynamic and non-monotonic. The logical framework introduced provides formal semantics of state descriptions of design processes based on (compositional) partial models and formal semantics of the reasoning behaviour based on (compositional) partial temporal models.

1. Introduction

In the area of diagnosis, a number of well-established logical theories have been developed and are acknowledged as valuable contributions to the field, such as Reiter (1987), and Console and Torasso (1990). For design (e.g., Brown and Chandrasekaran, 1989; French and Mostow, 1985; Logan, Corne, and Smithers, 1992; Takeda, Veerkamp, Tomiyama, and Yoshikawa, 1990) the situation differs. Although models for design have been proposed using logic as a vehicle (e.g., Coyne, 1988), and general design theories have been proposed (e.g., Tomiyama and Yoshikawa, 1987), formal semantics of both *static aspects* (i.e., characteristics of an individual state) of the design process and *dynamic aspects* (i.e., the reasoning behaviour) have yet to be defined. Design tasks typically reason with incomplete and inconsistent knowledge of requirements and design object descriptions: they reason non-monotonically with and about, for example, (default) assumptions, contradictory information, and new design knowledge. To handle such dynamic reasoning patterns, knowledge of tactics and strategies is needed. The formulation of the logical foundations (including formal semantics) of these patterns goes beyond classical logic.

In the current paper, a logical foundation is introduced in which formal semantics for both static and dynamic aspects of design are based on *partial models* (e.g., Langholm, 1988; Blamey, 1986). Partial models are a means to formalise *information states*, representing incomplete world descriptions (e.g., Langen and Treur, 1989); types of world descriptions relevant for design are design object descriptions and requirement sets. To obtain formal semantics of reasoning behaviour in design tasks, a recently developed approach based on partial temporal models is adopted, which has shown to be applicable to different types of (non-monotonic) reasoning (see Engelfriet and Treur, 1994; Gavrila and Treur, 1994; Treur, 1994). Semantics of a reasoning process is formalised by a set of (alternative) *reasoning traces*, represented by a *partial temporal model*, i.e., a sequence of partial models. As the partial models representing information states are used to provide semantics of the static aspects, a structural connection between the semantics of static aspects and of dynamic aspects is obtained. In Brazier, Langen, Ruttkay, and Treur (1994), the static aspects of design have been formalised; this paper elaborates upon that work with a formalisation of the dynamic aspects. Since a design task is complex (involving integration of different views and perspectives, and often different agents) and consists of a number of subtasks, the information states also have a compositional structure.

In Brazier, Langen, Ruttkay, and Treur (1994), an approach is presented to the development of intelligent design support systems based on a high-level formal specification language, as well as a generic task model of design specified in this formal specification language. This generic task model has been developed on the basis of the analysis of task models for the development of design support systems (e.g., Brumsen, Pannekeet and Treur, 1992; Geelen and Kowalczyk, 1992) and has been employed for the development of new design support systems (e.g., Brazier, Langen, Treur, Willems and Wijngaards, 1994). The logical foundations presented in this paper provide formal semantics for both the static and dynamic aspects of design tasks modelled by the generic task model. Moreover, the logical foundations can be used to establish (and prove) properties of design support systems, such as consistency, correctness, and completeness (see Treur and Willems, 1994a; 1994b), and to develop automated tools to support verification and validation of these properties. In most current frameworks such properties are basically static: they do not refer to the behaviour of the system. However, in interactive systems dynamic properties are also important (e.g., if under certain circumstances a particular type of behaviour of the system has been rejected by the user in the past, it should not be repeated in the present). Expressing dynamic properties requires a logical foundation for reasoning behaviour of a system (in interaction with the user)—this paper proposes a framework for this purpose.

In Section 2 of this paper, the notions of design process and design space are explained. In Sections 3 and 4, static and dynamic aspects of design processes are presented, respectively. In Section 5, an example of a design process is given. In Section 6, the logical theory of design is discussed and conclusions are drawn.

2. Design Processes and the Design Space

In design, requirement qualification sets and design object descriptions are manipulated. *Requirement qualifications* are qualitative expressions of the extent to which (individual or groups of) requirements must be met, either in isolation or in relation to each other. A *design object description* is a specification of the object to be created. A *design process* is described by a sequence of design decisions (and their rationale) concerning modifications to sets of requirements and their qualifications and to (partial) design object descriptions.

Figure 1 shows an example of a design process in the two-dimensional *design space* spanned by requirement qualification sets and design object descriptions. Note that the notions of space and dimension are used informally here: the choice of metric on the design space is left open. (A possibility would be to measure the distance between two points in a dimension by the number of differences between the descriptions denoted by these points.)

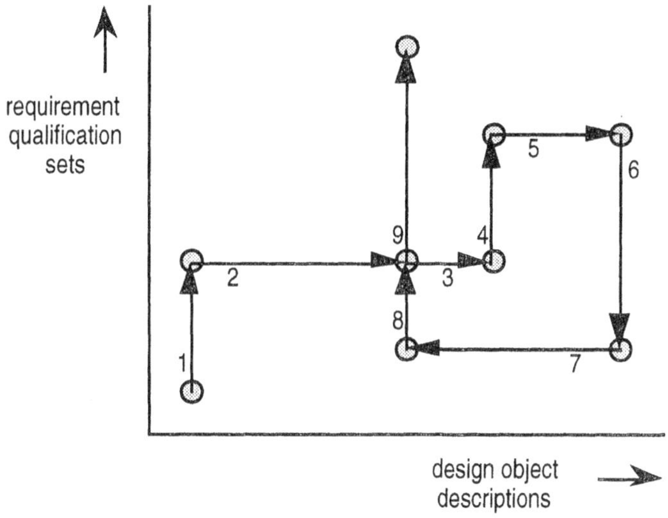

Figure 1. Example of a design process in the design space.

Figure 1 shows a nine-step sequence of modifications to requirement qualification set and design object description. The first step in the sequence, depicted by the arrow labelled '1', represents a modification to the initial requirement qualification set only. The initial design object description is modified in steps 2 and 3. After step 3 (i.e., at the point in which the arrow labelled '3' ends), modification of the design object description halts for some reason: maybe the design object description satisfies all requirements of the current requirement qualification set, or maybe there is reason to believe that no design object description can be made that satisfies all requirements. In each case, the requirement qualification set is modified in step 4, taking into account the reason why modification of the design object description stopped. After the modifications in steps 5 to 8, the design process reaches an interesting state: the sequence of modifications has led to a requirement qualification set and a design object description that in combination are equivalent to the result of step 2. Therefore another direction is sought, which differs from the one chosen in step 3, leading in step 9 to a modification to the requirement qualification set. In summary, there are five requirement qualification set modifications (steps 1, 4, 6, 8 and 9) and four design object description modifications (steps 2, 3, 5 and 7).

In general, a large (and possibly infinite) number of new points in the design space could be generated by modification to either the requirement qualification set or the design object description that correspond to a given point. In practice, only a few of these new points are of interest, because they are the ones that 'make sense'—these are the possible alternative choices for the next step in the design process. To describe the dynamics of the design process, knowledge of tactics and strategies, needed to guide the design process, must be made explicit.

3. Static Aspects of Design Processes

In design, manipulation of requirement sets, of their qualifications, and of design object descriptions plays a crucial role. A design process can be regarded as a sequence of design decisions concerning requirements, their qualifications and (partial) design object descriptions. The current state of the design process changes continually: requirements can be added or withdrawn, requirement qualifications can be changed, and partial design object descriptions can be added or retracted. During design, often different (alternative) requirement sets (and their qualifications) and design object descriptions are considered.

Steps in the design process can be represented by transitions of two types: transitions modifying design object descriptions and those modifying requirement qualification sets. Note that no commitment is made to model design as a search process. Steps in the design process can be controlled completely, depending on the strategic knowledge used.

The logical analysis of the static aspects of design processes is discussed below in Sections 3.2, 3.3, and 3.4. In Section 3.1, the basic terminology employed is introduced. Throughout Section 3 and Section 4, the example of designing a house will be used to illustrate the logical analysis.

3.1. BASIC APPROACH AND TERMINOLOGY ON STATIC ASPECTS

It is assumed that the reader is familiar with many-sorted first-order predicate logic, an essential element in our approach. In Langen and Treur, (1989), formal definitions of semantics for many-sorted partial models are presented. For an overview of partial logic, see Blamey (1986) and Langholm (1988).

Definition. A *signature* for many-sorted first-order predicate logic is a tuple $\Sigma = (\mathbf{S}, \mathbf{C}, \mathbf{F}, \mathbf{P})$ with sorts $\mathbf{S}$, constants $\mathbf{C}$, functions $\mathbf{F}$ and predicates $\mathbf{P}$.

In the sequel, Σ denotes a signature, $At(\Sigma)$ the set of all ground atoms of Σ and $Wff(\Sigma)$ the set of all closed well-formed formulae over Σ.

Definition. A *partial model* for Σ is a mapping $M: At(\Sigma) \rightarrow \{ 0, 1, u \}$. An atom $a \in At(\Sigma)$ is *true* in M if $M(a) = 1$, *false* in M if $M(a) = 0$, and *undefined* or *unknown* in M if $M(a) = u$. A partial model M is *complete* if for all $a \in At(\Sigma)$, either $M(a) = 0$ or $M(a) = 1$.

Definition. The *model space* $Mod(\Sigma)$ is the set of all partial models for Σ.

Definition. The *satisfaction relation* $\models$ on $Mod(\Sigma) \times Wff(\Sigma)$ is defined for all atomic well-formed formulae $a \in At(\Sigma)$ as:

$$M \models^{+} a \text{ iff } M(a) = 1$$
$$M \models^{-} a \text{ iff } M(a) = 0$$
$$M \not\models^{+} a \text{ iff } M(a) \neq 1$$
$$M \not\models^{-} a \text{ iff } M(a) \neq 0.$$

For the logical connectives $\neg, \wedge, \vee, \Rightarrow$ and $\Leftrightarrow$, the strong Kleene semantics is adopted (Blamey, 1986; Langholm, 1988), of which the truth tables are shown in Figure 2.

Definition. The *refinement relation* $\leq$ on $Mod(\Sigma) \times Mod(\Sigma)$ is such that for all $M, M' \in Mod(\Sigma)$, $M \leq M'$ holds if for all $a \in At(\Sigma)$, $M(a) \leq M'(a)$ (with $0 \leq 0, u \leq 0, 1 \leq 1, u \leq 1, u \leq u$).

Definition. A *theory* for Σ is a set $T \subset Wff(\Sigma)$.

Definition. Let M be a partial model for Σ and T a theory for Σ. The *class of models defined by* M *with* T is the set $\{ N \in Mod(\Sigma) \mid N \text{ is complete}, M \leq N, \text{and } N \models T \}$.

$\neg A$	
1	0
0	1
u	u

$A \wedge B$	1	0	u
1	1	0	u
0	0	0	0
u	u	0	u

$A \vee B$	1	0	u
1	1	1	1
0	1	0	u
u	1	u	u

$A \Rightarrow B$	1	0	u
1	1	0	u
0	1	1	1
u	1	u	u

$A \Leftrightarrow B$	1	0	u
1	1	0	u
0	0	1	u
u	u	u	u

Figure 2. Kleene's strong three-valued connectives.

3.2. STATIC ASPECTS OF DESIGN OBJECT DESCRIPTIONS

To describe a design object, a language is needed in which properties and their values can be named and relations between properties can be expressed.

Definition. A *design object description lexicon* is a signature $\Sigma_{DOD} = (\mathbf{S}, \mathbf{C}, \mathbf{F}, \mathbf{P})$, where $\{\text{Parameters}, \text{Values}\} \subseteq \mathbf{S}$ and $\{\text{eq} \subset \text{Parameters} \times \text{Values}\} \subseteq \mathbf{P}$.

In other words, the signature should at least contain the sorts Parameters and Values for denoting design parameters and values, respectively, and should contain a relation eq on Parameters $\times$ Values for denoting the fact that a certain design parameter has a certain value. Common relations used in design object ontologies, such as the 'part-of' relation by means of which the components and parts of a design object can be described, can also be included.

During design, not all properties of a design object are considered simultaneously: the description of a design object is often partial. In the sequel, Σ_{DOD} denotes a design object description lexicon.

Definition. A *design object description* based on Σ_{DOD} is a partial model for Σ_{DOD}.

Example. Suppose the designer has placed the living-room and the kitchen on the ground-floor (floor 0) and one bedroom (bedroom 1) on the first floor (floor 1). Whether the first bathroom (bathroom 1) and the second bedroom (bedroom 2) should be placed on the ground-floor or on the first floor is, as yet, undecided. This can be expressed by means of the following design object description DOD_1:

$$\mathrm{DOD}_1(\text{eq(floor-of(living-room(1)), 0)}) = 1$$
$$\mathrm{DOD}_1(\text{eq(floor-of(kitchen(1)), 0)}) = 1$$
$$\mathrm{DOD}_1(\text{eq(floor-of(bedroom(1)), 1)}) = 1$$
$$\mathrm{DOD}_1(\text{eq(floor-of(bedroom(2)), 1)}) = \mathrm{u}$$
$$\mathrm{DOD}_1(\text{eq(floor-of(bathroom(1)), 0)}) = \mathrm{u}$$
$$\mathrm{DOD}_1(\text{eq(floor-of(bathroom(1)), 1)}) = \mathrm{u}.$$

An abbreviated notation for DOD_1 is:

{ eq(floor-of(living-room(1)), 0),
eq(floor-of(kitchen(1)), 0),
eq(floor-of(bedroom(1)), 1) }.

A design object description can be seen as one element of the set of all partial or complete design object descriptions.

Definition. The *design object description space* DOD based on Σ_{DOD} is the set of tuples of models from $\mathrm{Mod}(\Sigma_{\mathrm{DOD}})$.

During design the number of properties of the design object that have been determined may increase or decrease. Both types of modification can be described by means of the following refinement relation.

Definition. The *design object refinement relation* based on Σ_{DOD} is the refinement relation $\leq$ on $\mathrm{Mod}(\Sigma_{\mathrm{DOD}}) \times \mathrm{Mod}(\Sigma_{\mathrm{DOD}})$. Furthermore, for any two tuples S and T that are elements of DOD, the *combined design object refinement relation* $S \leq T$ holds if for all $M \in S$ there is an $N \in T$ with $M \leq N$ and for all $N \in T$ there is an $M \in S$ with $M \leq N$.

Example. Suppose the designer, after allocating rooms as in the previous example, designs a kitchen with an area of 9 m^2. This can be expressed by means of the design object description DOD_2:

{ eq(floor-of(living-room(1)), 0), eq(floor-of(kitchen(1)), 0),
eq(floor-of(bedroom(1)), 1), eq(area(kitchen(1), m^2), 9) }.

Then $\mathrm{DOD}_1 \leq \mathrm{DOD}_2$.

To design an object, knowledge of the properties and the relations between the properties is essential. In practice, not all knowledge is available and has to be acquired during the design process.

Definition. A *design object theory* based on Σ_{DOD} is a theory T_{DOD} for Σ_{DOD}.

Example. Assume that general house design knowledge is that the area of a floor equals the sum of the areas of the rooms on that floor, and that a room can be allocated to one floor only. This can be expressed by means of the design object theory T'_{DOD}:

$$\forall f \in \text{Floors}\ \forall u \in \text{AreaUnits:}$$
$$\text{eq(area(f, u)),}$$
$$\sum a \in \text{Areas: } \exists r \in \text{Rooms (eq(floor-of(r), f)} \wedge \text{eq(area(r, u), a))}$$
$$\forall r \in \text{Rooms}\ \forall f_1, f_2 \in \text{Floors:}$$
$$\text{(eq(floor-of(r), } f_1) \wedge \text{eq(floor-of(r), } f_2)) \Rightarrow f_1 = f_2.$$

($\sum k$: $\varphi(k)$ is the sum over each k satisfying φ, such that if k satisfies exactly N quantifier-free instances of φ, then k appears exactly N times in the sum. The symbol '=' is the symmetric, reflexive and transitive equality relation on values.)

3.3. STATIC ASPECTS OF REQUIREMENT QUALIFICATION SETS

Before and during the process of design, knowledge of necessary and desired properties of the object to be designed (within a given context) is of importance. These necessary and desired properties are the requirements placed upon a design.

Definition. Let Σ_{DOD} be a design object description lexicon. A *requirement* is a well-formed formula over Σ_{DOD}.

To describe requirements, a language is needed in which requirements, qualifications and relations between qualifications can be expressed.

Definition. A *requirement qualification lexicon* is an extension of the signature Σ_{RQS} = (**S**, **C**, **F**, **P**), where

S : Sorts,	/* sorts in Σ_{DOD} */
Vars,	/* variables over Σ_{DOD} */

```
    VarSets,              /* variable sets over ΣDOD */
    Wffs,                 /* well-formed formulae over ΣDOD */
    WffTuples,            /* well-formed formula tuples over ΣDOD */
    QualificationNames,   /* names for qualifications */
    Parameters,           /* design parameters */
    Values;               /* values of design parameters */

C : Λ: WffTuples;         /* the empty tuple */
    ∅: VarSets;           /* the empty set */

F : ⟨ , ⟩: Wffs × WffTuples → WffTuples;      /* written as ⟨ , …, ⟩ */
    { , }: Vars × VarSets → VarSets;          /* written as { , …, } */
    eq: Parameters × Values → Wffs;
    and, or, implies: Wffs × Wffs → Wffs;
    not: Wffs → Wffs;
    for-all, exists: VarSets × Sorts × Wffs → Wffs;

P : rq ⊂ WffTuples × QualificationNames;  /* requirement qualification */.
```

The meaning of the above functions representing logical connectives is intuitive; see (Langen and Treur, 1989) for a definition. In the sequel, Σ_{RQS} denotes a requirement qualification lexicon.

Definition. A *requirement qualification set* based on Σ_{RQS} is a partial model for Σ_{RQS}.

Example. The customer's requirements for the design of the house are that: (1) there must always be a bathroom on the same floor as a bedroom, (2) the house has one kitchen, one living-room, three bedrooms of which one is on the ground-floor, and one bathroom and (3) the ground-floor area is at most 36 m^2. These are all hard requirements, i.e., a design must satisfy them all. This can be expressed by means of the requirement qualification set RQS_1:

```
{ rq(⟨for-all({f}, Floors, for-all({n}, RoomNrs,
        implies(eq(floor-of(bedroom(n)), f),
         exists({m}, RoomNrs, eq(floor-of(bathroom(m)), f))))))⟩, hard),
  rq(⟨exists({f}, Floors, eq(floor-of(kitchen(1)), f))⟩, hard),
  rq(⟨for-all({n}, RoomNrs,
       implies(exists({f}, Floors, eq(floor-of(kitchen(n)), f)), n=1))⟩, hard),
  rq(⟨exists({f}, Floors, eq(floor-of(living-room(1)), f))⟩, hard),
```

```
rq(⟨for-all({n}, RoomNrs,
     implies(exists({f}, Floors, eq(floor-of(living-room(n)), f)), n=1))⟩,
   hard),
rq(⟨eq(floor-of(bedroom(1)), 0)⟩, hard),
rq(⟨exists({f}, Floors, eq(floor-of(bedroom(2)), f))⟩, hard),
rq(⟨exists({f}, Floors, eq(floor-of(bedroom(3)), f))⟩, hard),
rq(⟨for-all({n}, RoomNrs,
     implies(exists({f}, Floors, eq(floor-of(bedroom(n)), f)),
             and(1≤n, n≤3)))⟩,
   hard),
rq(⟨exists({f}, Floors, eq(floor-of(bathroom(1)), f))⟩, hard),
rq(⟨for-all({n}, RoomNrs,
     implies(exists({f}, Floors, eq(floor-of(bathroom(n)), f)), n=1))⟩, hard),
rq(⟨for-all({a}, Areas,
     implies(eq(area(floor(0), m^2), a), ge(36, a))))⟩, hard) }
```

(where ge denotes the relation 'greater than or equal to').

Definition. The *requirement qualifications space* RQS based on Σ_{RQS} is the set of tuples from Mod(Σ_{RQS}).

Comparison of requirement qualification sets is necessary to guide the design process: knowledge is required of how qualifications are related and what the implications of the relations are.

Definition. A *requirement qualification theory* based on Σ_{RQS} is a theory T_{RQS} for Σ_{RQS}.

Example. Suppose that general building requirements require that there must be a hall on the ground-floor and that the minimum area of (1) a hall is 2 m^2, (2) a kitchen is 4 m^2, (3) a living-room is 16 m^2, (4) a bathroom is 3 m^2, and (5) a bedroom is 6 m^2. Furthermore, in general if the customer wants the kitchen on the ground-floor, then an additional requirement is that the living-room also be on the ground-floor. These hard requirements can be expressed by means of the requirement qualification theory T'_{RQS}:

```
rq(⟨exists({n}, RoomNrs, eq(floor-of(hall(n)), 0))⟩, hard)
rq(⟨for-all({n}, RoomNrs, for-all({a}, Areas,
     implies(eq(area(hall(n), m^2), a), ge(a, 2)))))⟩, hard)
rq(⟨for-all({n}, RoomNrs, for-all({a}, Areas,
     implies(eq(area(kitchen(n), m^2), a), ge(a, 4)))))⟩, hard)
```

rq(⟨for-all({n}, RoomNrs, for-all({a}, Areas,
implies(eq(area(living-room(n), m^2), a), ge(a, 16))))⟩, hard)
rq(⟨for-all({n}, RoomNrs, for-all({a}, Areas,
implies(eq(area(bathroom(n), m^2), a), ge(a, 3)))))⟩, hard)
rq(⟨for-all({n}, RoomNrs, for-all({a}, Areas,
implies(eq(area(bedroom(n), m^2), a), ge(a, 6)))))⟩, hard)
$\forall$m,n $\in$ RoomNrs:
rq(⟨eq(floor-of(kitchen(m)), 0)⟩, hard) $\Rightarrow$
rq(⟨eq(floor-of(living-room(n)), 0)⟩, hard).

3.4. STATIC ASPECTS OF THE DESIGN PROCESS AS A WHOLE

Given a number of requirement qualifications, specific tactics and strategies can be chosen to guide the overall design process (when to reason about requirements and their qualifications and when to reason about design object descriptions). These tactics and strategies determine on which requirements the design process is to (possibly temporarily) focus: a commitment is made to satisfy these requirements. In the sequel, Σ_{DOD} denotes a design object description lexicon and Σ_{RQS} a requirement qualification lexicon.

Definition. A *commitment mapping* from Σ_{RQS} to Σ_{DOD} is a mapping of partial models in $\text{Mod}(\Sigma_{RQS})$ onto sets of well-formed formulae in $\text{Wff}(\Sigma_{DOD})$.

The qualifications placed on requirements may be comparable. If one set of requirement qualifications specifies precisely the same as another, but in addition specifies extra requirement qualifications, the first is seen as a specialisation of the second.

Definition. Let T_{DOD} be a design object theory based on Σ_{DOD} and commit a commitment mapping from Σ_{RQS} to Σ_{DOD}. The *requirement qualification specialisation relation* based on Σ_{RQS} with T_{DOD} and commit is a relation $\leq$ on $\text{Mod}(\Sigma_{RQS}) \times \text{Mod}(\Sigma_{RQS})$ such that for all rqs_1, $rqs_2 \in \text{Mod}(\Sigma_{RQS})$:

$rqs_1 \leq rqs_2$ if for all $dod \in \text{Mod}(\Sigma_{DOD})$ such that $dod \models T_{DOD}$,
$dod \models \text{commit}(rqs_2)$ implies $dod \models \text{commit}(rqs_1)$.

As in the design object description space, this refinement relation can be extended to the requirements qualification space RQS, consisting of tuples.

Example. Suppose the requirement qualification set RQS_1 is refined to RQS_2 by applying the requirement qualification theory T'_{RQS} to RQS_1.

Suppose further that for the design only hard requirements are taken into account. This commitment can be expressed by means of the following mapping Commit':

$$\forall \text{rqs} \in \text{Mod}(\Sigma_{RQS})\ \forall \text{wff} \in \text{Wffs}:$$
$$\text{rq}(\langle \text{wff} \rangle, \text{hard}) \in \text{rqs} \Rightarrow \text{Commit'}(\text{rqs}) \models \text{wff}.$$

Then $RQS_1 \leq RQS_2$ with T'_{DOD} and Commit'.

A design problem can be seen as a problem of generating a description or modifying an existing description of a design object, given a number of requirement qualifications.

Definition. A *design problem description* is a pair (dod, rqs) with dod $\in$ $\text{Mod}(\Sigma_{DOD})$ and rqs $\in$ $\text{Mod}(\Sigma_{RQS})$.

The solution to a design problem is a design object description which fulfils the requirements chosen and which complies with the knowledge of the domain.

Definition. Let dod_0 and dod be design object descriptions based on Σ_{DOD}, T_{DOD} a design object theory based on Σ_{DOD}, rqs a requirement qualification set based on Σ_{RQS} and commit a commitment mapping from Σ_{RQS} to Σ_{DOD}. dod is a *design solution* of the design problem description (dod_0, rqs) with T_{DOD} and commit if (1) $dod_0 \leq dod$, (2) the class of models defined by dod with T_{DOD} is non-empty, and (3) for each element dod' of that class, dod' $\models$ commit(rqs).

Example. The design object description DOD_2 is *not* a design solution of the design problem description (DOD_1, RQS_2) with T'_{DOD} and Commit'.

4. Dynamic Aspects of Design Processes

To describe the dynamic aspects of a design process, the circumstances under which specific choices are to be made must be specified in relation to the alternatives. Strategic and tactical knowledge is required to steer the design process: that is, to determine along which of the two dimensions of the design space the design process should continue, and to determine how to proceed.

Section 4.1 defines the general basic concepts underlying the formalisation of the dynamic aspects of design: information states, transitions between information states and traces generated by these transitions. In Section 4.2 the notion of information state is more specifically defined for the information states relevant for design: design object (description) states, requirement qualification (set) states and overall control states. In Section 4.3 the related transitions are defined and in Section 4.4 the reasoning traces (temporal models of design process behaviour) based on the transitions are presented.

4.1. BASIC APPROACH AND TERMINOLOGY ON DYNAMIC ASPECTS

To define the dynamic aspects of a design process, a notion of state is required. In our logical approach, a state is the current state of the information acquired or derived so far, including information about incompleteness or partiality of the design process information.

Definition. An *information state* for signature Σ is a (partial) model M for Σ. The set of all information states for signature Σ is denoted by $IS(\Sigma)$.

An information state formalised as a partial model reflects all ground literal conclusions that have been derived at a certain moment in time. This approach can also be used to model inference relations such as SLD resolution or chaining.

Definition. A *transition between information states* for signature Σ is a pair of partial models for Σ; i.e., an element $\langle s, s' \rangle$ of $IS(\Sigma) \times IS(\Sigma)$. A *transition relation* is defined as a set of transitions, i.e. a relation on $IS(\Sigma) \times IS(\Sigma)$. If this relation is defined as a mapping from $IS(\Sigma)$ into $IS(\Sigma)$, it is called a *transition function.*

Definition. A *trace* or *partial temporal model* for signature Σ is a sequence of information states $(M^t)_{t \in \mathbb{N}}$ in $IS(\Sigma)$. The set of all partial temporal models is denoted by $IS(\Sigma)^{\mathbb{N}}$, or $Traces(\Sigma)$.

Traces generated by repeatedly applying a transition function on the current information state can be interpreted as partial temporal models. These partial temporal models provide a declarative description of the semantics of the behaviour of the design process; the set of these models can be viewed as the required behaviour of the design process.

If a design process is modelled as a compositional structure, then the information state is a combination of information (sub-)states of each of the components of the structure. Transitions from one information state to another are specified in a similar way by their effect on the different information sub states. The overall partial temporal model, that models the behaviour of the design process, can be constructed as a composition of partial temporal models of each of the components.

4.2. STATES IN A DESIGN PROCESS

An information state of the design process comprises information on a design object description and a requirement qualification set. The abbreviations used below are DOD for design object description space and RQS for requirement qualification set space.

Definition. (design object states and requirement qualification states)
a) A *design object state* is an element of $IS_{DOD} = IS_{DOD}{}^{object} \times IS_{DOD}{}^{meta}$, where $IS_{DOD}{}^{object} = IS(\Sigma_{DOD}{}^{object})$ and $IS_{DOD}{}^{meta} = IS(\Sigma_{DOD}{}^{meta})$, with $\Sigma_{DOD}{}^{object}$ and $\Sigma_{DOD}{}^{meta}$ signatures for the object-information and meta-information about design object descriptions, respectively.
b) A *requirement qualification state* is an element of $IS_{RQS} = IS_{RQS}{}^{object} \times IS_{RQS}{}^{meta}$, where $IS_{RQS}{}^{object} = IS(\Sigma_{RQS}{}^{object})$ and $IS_{RQS}{}^{meta} = IS(\Sigma_{RQS}{}^{meta})$, with $\Sigma_{RQS}{}^{object}$ and $\Sigma_{RQS}{}^{meta}$ signatures for the object information and meta-information about requirement qualification sets, respectively.

Example. The designer often needs to reason at a meta-level about a partial design object description, for instance with respect to completeness. For example (cf. DOD_1), the designer knows that the living-room and the kitchen are on the ground-floor and not on the first floor and that the first bedroom is on the first floor and not on the ground-floor. In addition, he/she knows that the floor for a second bedroom has not yet been decided. This can be expressed by means of the following design object state IS'_{DOD}:

```
⟨ { eq(floor-of(living-room), 0),
    eq(floor-of(kitchen), 0),
    eq(floor-of(bedroom(1)), 1) },
  { true(eq(floor-of(living-room), 0)),
    false(eq(floor-of(living-room), 1)),
    true(eq(floor-of(kitchen), 0)),
    false(eq(floor-of(kitchen), 1)),
    true(eq(floor-of(bedroom(1)), 1)),
    false(eq(floor-of(bedroom(1)), 0)),
    ¬ known(eq(floor-of(bedroom(2)), 0)),
    ¬ known(eq(floor-of(bedroom(2)), 1))} ⟩.
```

In a similar way, a requirement qualification state IS'_{RQS} can be defined as a pair $\langle s_{object}, s_{meta} \rangle$, where s_{object} equals, for example, the requirement qualification set RQS_1 (cf. Section 2.3) and s_{meta} comprises the meta-information about RQS_1 that all requirement qualifications in RQS_1 are known to be true.

As can been seen in the above example, one part of the meta-information about a design object description or a requirement qualification set concerns epistemic information (i.e., information about what is known). The full epistemic information IS_e associated with an object information state IS_o is:

$$IS_o(a) = 1 \Leftrightarrow (\ IS_e(true(a)) = 1 \wedge IS_e(false(a)) = 0 \wedge IS_e(known(a)) = 1\ )$$
$$IS_o(a) = 0 \Leftrightarrow (\ IS_e(true(a)) = 0 \wedge IS_e(false(a)) = 1 \wedge IS_e(known(a)) = 1\ )$$
$$IS_o(a) = u \Leftrightarrow (\ IS_e(true(a)) = 0 \wedge IS_e(false(a)) = 0 \wedge IS_e(known(a)) = 0\ ).$$

Besides epistemic information, the meta-information also includes local control information, which directs the design process within either the design object description space or the requirement qualifications space.

Overall design process coordination is needed to determine in which of these two spaces the design process is to continue. Therefore, a third state of design process coordination information is defined, expressed in terms taken from an overall control lexicon $\Sigma_{DS}{}^{control}$. For the design system, the abbreviation DS is used.

Definition. (states of a design process)
a) A *basic state of a design process* is a pair consisting of a design object state and a requirement qualification state, i.e., an element of $IS_{DS}{}^{basic} = IS_{DOD} \times IS_{RQS}$.
b) An *overall state of a design process* is a pair consisting of a basic state of the design process and an overall control state, i.e., an element of $IS_{DS}{}^{overall} = IS_{DS}{}^{basic} \times IS_{DS}{}^{control}$, where $IS_{DS}{}^{control} = IS(\Sigma_{DS}{}^{control})$.

4.3. DESIGN STEPS

Having defined states, design steps can be defined by transitions from one state to another. This can be described in the following compositional manner.

Definition. (transitions in the two spaces)
a) A *transition in the design object space* is a pair of design object states, i.e., an element of $IS_{DOD} \times IS_{DOD}$.
b) A *transition in the requirement qualification space* is a pair of requirement qualification states, i.e., an element of $IS_{RQS} \times IS_{RQS}$.

Definition. (basic and overall design transitions)
a) A *basic design transition* is a pair of basic design states, i.e., an element of $IS_{DS}{}^{basic} \times IS_{DS}{}^{basic}$ that is induced by a transition in either the design object space or the requirement qualification space.

b) An *overall control transition* is a pair of control states, i.e., an element of $IS_{DS}^{control} \times IS_{DS}^{control}$.
c) An *upward control interaction transition* is a pair consisting of a basic design state and a control state, i.e., an element of $IS_{DS}^{basic} \times IS_{DS}^{overall}$ that is induced by a transition in either the design object space or the requirement qualification space.
d) A *downward control interaction transition* is a pair consisting of a control state and a basic design state, i.e., an element of $IS_{DS}^{overall} \times IS_{DS}^{basic}$ that is induced by a transition in either the design object space or the requirement qualification space.
e) An *overall transition* is a pair consisting of two overall design states, i.e., an element of $IS_{DS}^{overall} \times IS_{DS}^{overall}$ that is induced by one of the above transition types.

For each of these types of transitions, it holds that if an individual transition is element of $S \times S'$, a *transition relation* of that type is defined as a subset of $S \times S'$. Furthermore, if this relation is defined as a mapping from S into S', it is called a *transition function.* It will be assumed that in upward and downward control interactions, only the meta-level information of the basic design states is involved. Examples of basic design transitions are shown in Section 5.

4.4. TRACES AND TEMPORAL MODELS OF A DESIGN PROCESS

Having defined states and transitions in a compositional manner, traces can be defined.

Definition. (overall temporal model) Let $Traces_{DS} = (IS_{DS}^{overall})^{\mathbb{N}}$. An *overall trace* is an element $(M^t)_{t \in \mathbb{N}} \in Traces_{DS}$. Such a trace $(M^t)_{t \in \mathbb{N}}$ is *a temporal model of a design system* if for all time points t the step from M^t to M^{t+1} is defined in accordance with an overall transition. The set $BehMod$ of temporal models forms a subset of $Traces_{DS}$.

A trace defines a complete design history. In most systems only part of the design history is actually represented (see for instance (Brazier, Langen, Treur, Willems, and Wijngaards, 1994), where it was sufficient for devising an elevator configuration to remember the previous state of the configuration). An overall temporal model describes a trace representing possible (intended) behaviour of the design process. From every initial information setting, traces can be generated by the transitions. All generated traces together form the set $BehMod$. The transition functions in fact define a set of (temporal) axioms $BehTheory$ on temporal models in $Traces_{DS}$. The possible behavioural alternatives are given by the set of the temporal models satisfying these temporal axioms. A design process is correct with respect to the specified transitions if each generated trace (from BehMod) satisfies the theory BehTheory. This can be used for purposes of verification or proving properties of a specification. For example, proof techniques in temporal logic can be used to derive whether a design system is able to generate a given design object description on the basis of a given set of requirement qualifications. For more details on verification, see (Treur and Willems, 1994a; 1994b).

5. Example of a Design Process

In this section, the example of designing a house is pursued to show an overall trace of a design process. In this example, a customer and a designer cooperate in the design: the customer by stating his/her wishes with regards to rooms, floors and room areas, and the designer by allocating rooms to floors and determining the areas of rooms.

The sample process proceeds as follows. First, the customer states his/her wishes, which are then translated into requirements and qualifications (cf. the set RQS_1 in Section 3.3). After this, the designer tries to design a bungalow that fulfils the requirements. This, however, results in a design with too large a ground-floor area. The designer cannot remedy this problem: adding one storey to the house and putting a bedroom on the first floor also entails putting a bathroom on that floor, but that would mean there would be more bathrooms than the customer wanted. To resolve this problem, the customer decides to allow for more than one bathroom. The designer then designs a two-storey house that pleases the customer.

A (partial) overall trace of this process is shown below. Of each element $(M^t)_{t \in \mathbb{N}}$ from this trace, the contents of its five components, $IS_{DOD}{}^{object}$, $IS_{DOD}{}^{meta}$, $IS_{RQS}{}^{object}$, $IS_{RQS}{}^{meta}$, and $IS_{DS}{}^{control}$, are shown. Together, these states (in chronological order) form the design history.

The requirement qualification sets that are generated during the design process are written as $RQS_{i \in \mathbb{N}}$, and similarly, the design object descriptions as $DOD_{j \in \mathbb{N}}$. Note that initially, $RQS_0 = \emptyset$ and $DOD_0 = \emptyset$. For the sake of convenience, the meta-information in states of $IS_{DOD}{}^{meta}$ and $IS_{RQS}{}^{meta}$ is restricted to the (partial) results of analysis of the corresponding object-information and the chosen method of modification. Similarly, the overall control information in states of $IS_{DS}{}^{control}$ is restricted to information about which description to be manipulated next and how.

Step 1. The customer states his/her wishes, which are translated into a set of requirements for the design of the house ('$\cup$' is the set union operation):

$RQS_1 = RQS_0 \cup$
{ /* there must always be a bathroom on the same floor as a bedroom */
rq(⟨for-all({f}, Floors, for-all({n}, RoomNrs,
implies(eq(floor-of(bedroom(n)), f),
exists({m}, RoomNrs, eq(floor-of(bathroom(m)), f))))))⟩, hard),

/* the house has one kitchen */
rq(⟨exists({f}, Floors, eq(floor-of(kitchen(1)), f))⟩, hard),
rq(⟨for-all({n}, RoomNrs,
implies(exists({f}, Floors, eq(floor-of(kitchen(n)), f)), n=1))⟩, hard),

/* the house has one living-room */
rq(⟨exists({f}, Floors, eq(floor-of(living-room(1)), f))⟩, hard),
rq(⟨for-all({n}, RoomNrs,
implies(exists({f}, Floors, eq(floor-of(living-room(n)), f)), n=1))⟩,
hard),

/* the house has three bedrooms of which one is on the ground-floor */
rq(⟨eq(floor-of(bedroom(1)), 0)⟩, hard),
rq(⟨exists({f}, Floors, eq(floor-of(bedroom(2)), f))⟩, hard),
rq(⟨exists({f}, Floors, eq(floor-of(bedroom(3)), f))⟩, hard),
rq(⟨for-all({n}, RoomNrs,
implies(exists({f}, Floors, eq(floor-of(bedroom(n)), f)),
and(1≤n, n≤3)))⟩,
hard),

/* the house has one bathroom */
rq(⟨exists({f}, Floors, eq(floor-of(bathroom(1)), f))⟩, hard),
rq(⟨for-all({n}, RoomNrs,
implies(exists({f}, Floors, eq(floor-of(bathroom(n)), f)), n=1))⟩, hard),

/* the ground-floor area is at most 36 m^2 */
rq(⟨for-all({a}, Areas,
implies(eq(area(floor(0), m^2), a), ge(36, a))))⟩, hard) }.

Step 2. The current requirement qualification set is analysed, and it is found that it can be further refined by extending it with all logical consequences that follow from the available theory of the domain (T'_{RQS}, Section 3.3):

{ analysis(current-description-can-be-refined),
method(deductive-refinement) }.

Step 3. The current requirement qualification set is deductively refined by means of T'_{RQS}:

$RQS_2 = RQS_1 \cup$
{ /* there must be a hall on the ground-floor */
rq(⟨exists({n}, RoomNrs, eq(floor-of(hall(n)), 0))⟩, hard),

/* the minimum area of a hall is 2 m^2 */
rq(⟨for-all({n}, RoomNrs, for-all({a}, Areas,
implies(eq(area(hall(n), m^2), a), ge(a, 2)))))⟩, hard),

/* the minimum area of a kitchen is 4 m^2 */
rq(⟨for-all({n}, RoomNrs, for-all({a}, Areas,
implies(eq(area(kitchen(n), m^2), a), ge(a, 4)))))⟩, hard),

/* the minimum area of a living-room is 16 m^2 */
rq(⟨for-all({n}, RoomNrs, for-all({a}, Areas,
implies(eq(area(living-room(n), m^2), a), ge(a, 16)))))⟩, hard),

/* the minimum area of a bathroom is 3 m^2 */
rq(⟨for-all({n}, RoomNrs, for-all({a}, Areas,
implies(eq(area(bathroom(n), m^2), a), ge(a, 3)))))⟩, hard),

/* the minimum area of a bedroom is 6 m^2 */
rq(⟨for-all({n}, RoomNrs, for-all({a}, Areas,
implies(eq(area(bedroom(n), m^2), a), ge(a, 6)))))⟩, hard).

Step 4. The current requirement qualification set is analysed and no further problems can be found:

{ ¬ analysis(current-description-can-be-refined),
¬ analysis(current-description-is-too-restrictive) }.

Step 5. The current design process is analysed, and it is determined that it is now time to refine the current design object description:

{ to-manipulate-next(current-design-object-description),
manipulation-type(refinement) }.

Step 6. The current design object description is analysed, and it is found that it is incomplete and should be refined by making assumptions about useful extensions to the current description:

{ analysis(current-description-is-incomplete),
method(refinement-by-assumptions) }.

Step 7. The designer's first idea is to design a bungalow, with a kitchen of 4 m^2, a living-room of 16 m^2, a hall of 2 m^2, a bathroom of 3 m^2, and three bedrooms, each of 6 m^2:

$DOD_1 = DOD_0 \cup$
{ eq(floor-of(kitchen(1)), 0),
eq(area(kitchen(1), m^2), 4),
eq(floor-of(living-room(1)), 0),
eq(area(living-room(1), m^2), 16),
eq(floor-of(hall(1)), 0),
eq(area(hall(1), m^2), 2),
eq(floor-of(bathroom(1)), 0),
eq(area(bathroom(1), m^2), 3),
eq(floor-of(bedroom(1)), 0),
eq(area(bedroom(1), m^2), 6),
eq(floor-of(bedroom(2)), 0),
eq(area(bedroom(2), m^2), 6),
eq(floor-of(bedroom(3)), 0),
eq(area(bedroom(3), m^2), 6) }.

Step 8. The current design object description is analysed, and it is found that it can be further refined by extending it with all logical consequences that follow from the available theory of the domain (T'_{DOD}, Section 3.2):

{ analysis(current-description-can-be-refined),
method(deductive-refinement) }.

Step 9. The current design object description is deductively refined by means of T'_{DOD}:

$DOD_2 = DOD_1 \cup$ { eq(area(floor(0), m^2), 43) }.

Step 10. The current design object description is analysed, and it is found that it is incorrect, because of a violation of requirements, in particular the requirement on the maximum floor area, and should therefore be revised:

{ analysis(current-description-is-incorrect),
method(revision) }.

Step 11. The designer understands that the idea of designing a bungalow is not so good, because the floor area will always remain a problem. Therefore, he/she now tries a two-storey house. The only difference with the bungalow design is that the two-storey house has two of the three bedrooms on the first floor rather than on the ground floor ('\' is the set difference operation):

$DOD_3 = DOD_1 \cup$
{ eq(floor-of(bedroom(2)), 1),
eq(floor-of(bedroom(3)), 1) }
\
{ eq(floor-of(bedroom(2)), 0),
eq(floor-of(bedroom(3)), 0) }.

Step 12. The current design object description is analysed, and it is found that it is still incorrect, because of a violation of requirements, in particular the requirement on the number of bathrooms in the house, and should therefore be revised:

{ analysis(current-description-is-incorrect),
method(revision) }.

Step 13. The designer does not know how to proceed: whatever he/she does, a violation of requirements seems unavoidable. Bedrooms on two floors also requires bathrooms on two floors, but there may only be one bathroom.

Step 14. The current design process is analysed, and it is determined that it is now time to manipulate the current requirement qualification set:

{ to-manipulate-next(current-requirement-qualification-set),
manipulation-type(revision) }.

Step 15. The current requirement qualification set is analysed, and it is found that it is too restrictive to permit any design solution, which can be resolved by deleting one or more requirement qualifications:

{ analysis(current-description-is-too-restrictive),
method(deletion)}.

Step 16. The customer, knowing the reason why the preliminary design of the two-storey house failed, drops the hard single-bathroom requirement:
$RQS_3 = RQS_2 \setminus$
{ rq(⟨for-all({n}, RoomNrs,

implies(exists({f}, Floors, eq(floor-of(bathroom(n)), f)), n=1))⟩,
hard) }.

Step 17. The current requirement qualification set is analysed and no further problems can be found:

{ ¬ analysis(current-description-can-be-refined),
¬ analysis(current-description-is-too-restrictive) }.

Step 18. The current design process is analysed, and it is determined that it is now time to revise the current design object description:

{ to-manipulate-next(current-design-object-description),
manipulation-type(revision) }.

Step 19. The current design object description is analysed, and it is found that it is (still) incorrect and should be revised:

{ analysis(current-description-is-incomplete),
method(revision) }.

Step 20. The designer proceeds with the design of the two-storey house and need not throw any parts away. The only thing he/she does is to place a bathroom on the first floor, with an area of 3 m^2:

$DOD_4 = DOD_3 \cup$
{ eq(floor-of(bathroom(2), 1), eq(area(bathroom(2), m^2), 3) }.

Step 21. The current design object description is analysed, and it is found that it can be further refined by extending it with all logical consequences that follow from the available theory of the domain (T'_{DOD}, Section 3.2):

{ analysis(current-description-can-be-refined),
method(deductive-refinement) }.

Step 22. The current design object description is deductively refined by means of T'_{DOD}:

$DOD_5 : DOD_4 \cup$
{ eq(area(floor(0), m^2), 31), eq(area(floor(1), m^2), 15) }.

Step 23. The current design object description is analysed and, since it is complete and satisfies all requirements, no more problems are found:

{ ¬ analysis(current-description-is-incorrect),
¬ analysis(current-description-is-incomplete) }.

6. Discussion and Conclusions

A logical framework, capturing both *static* and *dynamic* aspects of design has been presented in this paper. It constitutes a logical theory of design which can be (and has been) instantiated for different types of design tasks (cf. Geelen and Kowalczyk, 1992; Brumsen, Pannekeet, and Treur, 1992).

The formal analysis of the dynamic aspects of design processes provides an explicit means to model design strategies. Declarative specifications of strategies provide a basis for interaction between autonomous systems on, for example, the strategy employed during design. As expert designers often wish to determine the design strategy employed, flexibility is mandatory. By formally defining the strategies involved, design support systems can be designed within which the user is given the freedom to determine how a task is to be approached. Formal specifications, together with well-defined semantics, provide a basis for such flexibility and a basis for the verification and validation of design support systems' behaviour.

Current research focusses on fundamental issues with respect to the formalisation of design strategies, (non-monotonic) reasoning patterns, verification, validation and knowledge acquisition.

Acknowledgements

This research has been partially supported by the Dutch Foundation for Knowledge-Based Systems (SKBS) within the A3 project "An environment for modular knowledge-based systems (based on meta-knowledge) for design tasks." The constructive comments and suggestions for improvements provided by Tim Smithers have been much appreciated.

References

Blamey, S.: 1986, Partial logic, *in* D. Gabbay and F. Günthner (eds), *Handbook of Philosophical Logic*, Reidel, Dordrecht, III, pp. 1–70.

Brazier, F. M. T., Langen, P. H. G. van, Ruttkay, Zs., and Treur, J.: 1994, On formal specification of design tasks, *in* J. S. Gero and F. Sudweeks (eds), *Proceedings Artificial Intelligence in Design'94*, Kluwer, Dordrecht, pp. 535–552.

Brazier, F. M. T., Langen, P. H. G. van, Treur, J., Wijngaards, N. J. E., and Willems, M.: 1994, Modelling a design task in DESIRE: The VT example, *Technical Report IR-377*, Artificial Intelligence Group, Department of Mathematics and Computer Science, Vrije Universiteit, Amsterdam. Also in A. Th. Schreiber and W. Birmingham (eds) (1995), *International Journal on Human-Computer Studies, Special Issue on Sisyphus.*

Brown, D. C. and Chandrasekaran, B.: 1989, *Design Problem Solving: Knowledge Structures and Control Strategies*, Pitman, London.

Brumsen, H. A., Pannekeet, J. H. M., and Treur, J.: 1992, A compositional knowledge-based architecture modelling process aspects of design tasks, *Proceedings of the Twelfth International Conference on Artificial Intelligence, Expert Systems and Natural Language (Avignon-92)*, EC2, Nanterre, Vol. 1, pp. 283–294.

Console, L., and Torasso, P.: 1990, Hypothetical reasoning in causal models, *International Journal of Intelligent Systems*, **5**(1), 83–124.

Coyne, R. D.: 1988, *Logic Models of Design*, Pitman, London.

Engelfriet, J. and Treur, J.: 1994, Temporal theories of reasoning, *Proceedings of the Fourth European Workshop on Logics in Artificial Intelligence (JELIA'94)*, Springer-Verlag, Berlin.

French, R. and Mostow, J.: 1985, Toward better models of the design process, *AI Magazine*, **6**(1), 44–57.

Gavrila, I. S. and Treur, J.: 1994, A formal model for the dynamics of compositional reasoning systems, *in* A. G. Cohn (ed.), *Proceedings of the Eleventh European Conference on Artificial Intelligence (ECAI '94)*, John Wiley, Chichester, pp. 307–311.

Geelen, P. A. and Kowalczyk, W.: 1992, A knowledge-based system for the routing of international blank payment orders, *Proceedings Twelfth International Conference on Artificial Intelligence, Expert Systems and Natural Language (Avignon-92)*, EC2, Nanterre, Vol. 2, pp. 669–677.

Langen, P. H. G. van, and Treur, J.: 1989, Representing World Situations and Information States by Many-Sorted Partial Models, *Technical Report PE8904*, Programming Research Group, Department of Mathematics and Computer Science, University of Amsterdam, Amsterdam.

Langholm, T.: 1988, Partiality, Truth and Persistence, *CSLI Lecture Notes No. 15*, Stanford University, Stanford, CA.

Logan, B. S., Corne, D. W., and Smithers, T.: 1992, Enduring support: on defeasible reasoning in design support systems, *in* J. S. Gero (ed.), *Artificial Intelligence in Design '92*, Kluwer, Dordrecht, pp. 433–454.

Reiter, R.: 1987, A theory of diagnosis from first principles, *Artificial Intelligence*, **32**, 57–95.

Takeda, H., Veerkamp, P. J., Tomiyama, T. and Yoshikawa, H.: 1990, Modelling design processes, *AI Magazine*, **11**(4), 37–48.

Tomiyama, T. and Yoshikawa, H.: 1987, Extended general design theory, *in* H. Yoshikawa and E. A. Warman (eds), *Proceedings IFIP WG 5.2 Working Conference on Design Theory for CAD*, North-Holland, Amsterdam, pp. 95–125.

Treur, J.: 1991, A logical framework for design processes, *in* P. J. W. ten Hagen and P. J. Veerkamp (eds), *Intelligent CAD Systems III, Proceedings of the Third Eurographics Workshop on Intelligent CAD Systems*, Springer-Verlag, Berlin, pp. 3–20.

Treur, J.: 1994, Temporal semantics of meta-level architectures for dynamic control of reasoning, *Proceedings of the Fourth International Workshop on Meta-Programming in Logic (META '94)*, Springer-Verlag, Berlin, Lecture Notes in Computer Science 883.

Treur, J., and Willems, M.: 1994a, A logical foundation for verification, *in* A. G. Cohn (ed.), *Proceedings of the Eleventh European Conference on Artificial Intelligence (ECAI '94)*, John Wiley, Chichester, pp. 745–749.

Treur, J., and Willems, M.: 1994b, On verification in compositional knowledge-based systems, *in* A. Preece (ed.), *Proceedings of the ECAI '94 Workshop on Validation of Knowledge-Based Systems*, Amsterdam, pp. 4–20. *Also*: Formal notions for verification of dynamics of knowledge-based systems, *in* M.-C. Rousset and M. Ayel (eds.) (1995), *Proceedings of the European Symposium on Validation and Verification of KBSs (EUROVAV '95)*, Chambéry.

15

REPRESENTING THE COLLABORATIVE DESIGN PROCESS: A PRODUCT MODEL-ORIENTED APPROACH

BANGYU LEI, TOSHIHARU TAURA
The University of Tokyo, Japan

AND

JUN NUMATA
SONY Systems Design Co., Ltd, Japan

Abstract. The collaborative design process can be viewed essentially as the evolution of product data, which is the results of a series of decisions. Contemporary product data representations in describing design deliberations are either informal with plain text or insufficient using some simple data structures. Explicit modeling of the evolution, alternatives and constraints of the product data in a large design space is crucial for capturing the process information at any a state of its recorded history. This paper develops a product data model on the basis of the integrated generic resources from STEP[1], which is used to formally describe the objective of a task, assignments and alternatives etc. of a decision, and in particulars, the constraints among decisions. A product model-oriented representation of the collaborative design process is proposed to develop a database which addresses the archiving of design history. The proposed representation focuses on the formal specifications of the process ingredients and the dependencies among these ingredients, such as product-data-model, assignment, activity, task, negotiation, and agent. Accordingly, a data model is developed in EXPRESS[2]. An object-oriented database is under development to implement the data model .

1. Introduction

The attitude of designers describing a design process is similar to that of mathematicians describing a theorem-proving process. In the formal design documentation such as drawings and design reports, the painstaking processes of trial and error, revision and conflict adjustment are all invisible. Designs are revamped and polished until all traces of how they were developed are completely hidden (Banares-Alcantara, 1991). For this reason, it may be more productive to focus on the results of the designer's thinking process rather than to address the thinking process itself. In other words, it is

[1]Standard for the Exchange of Product Model Data, -- ISO 10303.
[2]A formal data specification language, specified in ISO 10303-11.

much easier to observe the events that transpired during the process. The design activities which take place when a state transition of the artifact being designed is made will provide important information. Recording all possible solutions and the argumentation for each decision-making provides access to the alternatives that were identified and the reasons for selecting or rejecting them. By making this information and its evolution explicit, other designers can avoid considering the same unfruitful areas or find some new chances when modification, redesign, or new design in the future. A number of researchers have hypothesized that capture and reuse of this evolving information has the potential for improving the design process and reusing of design information (Ullman, 1994).

On the other hand, the complexity of modern designing often demands integrated design teams. In industry, frequently, the domain knowledge required to develop a product is available; the requirements, constraints and primitive design components are usually established over decades of practices; however, the limiting factor determining the speed of product development is the efficiency with which the information environment can be coordinated to develop the product. This results in growing demands on capture and representation of not only the rationale behind design decisions of individual designers for a shared understanding, but also the coordination activities of multiple members of a design team within multiple tasks, which bind the information accessed, shared and generated, during the collaborative design process.

Recently, there is increasing interest in capturing information about design processes and recording the rationale behind the decisions affecting the evolution of the information. Published work has used the terms "design history", "design rationale", and "design intent" that manage the capture, storage and query of the evolving information (Ullman, 1994; Brown, 1994; Ganeshan et al., 1994; Chung and Goodwin, 1994). However, this effort is still in its infancy and gives rise to many questions on modeling and controlling design information in the design process, especially in the collaborative design process.

From the viewpoint of the artificial intelligence community, design has the following properties. (1) Design is an opportunistic activity. It is not performed using a fixed set of operators applied in an ordered way. Design is a process wherein various design activities occur in an opportunistic manner, either top-down or bottom-up. (2) Design is an exploration activity. It is classified as exploration (Smithers et al., 1989) rather than search, because knowledge about the space of possible solutions has to be obtained before goals can be well formulated. Typically, the initial description of the solution is incomplete and/or ambiguous and/or inconsistent. Design has a large space of possible solutions. Complexity arises from the very large number of alternatives of intermediate and final designs. (3) Problems of consistency are inevitable. Since design is an incremental activity, it is certain to reverse some of the previous decisions either for refinement, or for resolution of conflicts with other decisions self-made or made by other team members. Maintaining consistency among multiple decisions is one of the characteristics of the collaborative design process.

Taking these properties into full account, we present a product model-oriented representation of the collaborative design process, to build a data model for a database which addresses the archiving of design history and serves as the information infrastructure of a concurrent design environment in the context of an industry project at Sony Corporation. The proposed representation regards the collaborative design process as the evolution of product data. The product evolution can be viewed as the result of a series of decisions, i.e., *activities* and *negotiations*, each of which is one step in the transformation of the *product*, made by multiple *design agents*. A STEP-based (ISO, 1993a) *product data model* is developed to support the description of *objectives* of a task, *assignments* and *alternatives* of a decision, and so on. Design *activities* and *negotiations* are organized into *tasks* which are charged by *agents*. The *product data model* plays a key role in task decomposition and the interactions between the agents by providing a common ontology on product data access and consistency maintenance.

In the next section, previous approaches taken for design history representation are reviewed. The third section introduces a product data model on the basis of the integrated generic resources from STEP. Then, ingredients representing the collaborative design process are formalized and specified in EXPRESS (ISO, 1993b) using the product data model. The discussion of the current application state and desired future developments of the proposed approach, and conclusions follow.

2. Related Work

Since Mostow (1985) stated that there is a growing consensus in the artificial intelligence community that "An idealized design history is a useful abstraction of the design process," both the artificial intelligence and design science communities have been active in developing the concept of the design history as records of the rationale behind design decisions and of the intent of the designers. The root of most of the previous work lies in the IBIS[3] method (Rittel et al., 1973) for policy decision-making in the domain of government administration and planning where the deliberation process for complex problems is viewed as a process of negotiation among different groups with different stakes in the problem in terms of *issues* (tasks,

[3]Issue Based Information System.

questions or problems), *alternatives* (proposals or concepts), *arguments* (evaluations) and *decisions*.

In the artificial intelligence community, previous work focused on the development of IBIS-based tools, e.g., gIBIS, Potts and Brun model, DRL[4], and DRCS. gIBIS is a computer tool to capture design histories and support computer-mediated teamwork (Conklin and Begeman, 1988). Potts and Brun (1988) distinguish between two types of design information: the process of design and the product of design. Designers work from an initial design problem to the final design by identifying alternatives, exploring them and then selecting one that satisfies or moves towards satisfying the design objectives. Lee (1991) extended Potts and Brun's model to develop DRL. In most embodiments of IBIS, artifact information such as goals, alternatives and specifications has been left informal. Plain textual representations are used for describing design deliberations. These efforts provide a means of organizing these deliberations in the form of nodes and links within the computer. In contrast to the natural language text representation for the contents of the network's nodes, a structured language, DRCS, attempts to represent the product and the process (Klein 1993). However, it is quite general and still under development and untested (Ullman, 1994).

The design science community has introduced the IBIS method to address the capture of design histories in different fields such as mechanical design (Ullman, 1994; Brown, 1994; Nagy and Ullman, 1992; Thompson and Lu, 1990; Chen et al., 1990), civil design (Ganeshan et al., 1994; Rosenman et al., 1994) and chemical plant design (Chung and Goodwin, 1994). Researchers in this community observe the limitations of IBIS-based approaches developed by the artificial intelligence people, that is, informal plain text representation for product, and implicit description of the constraints among decisions. To overcome these shortcomings, researchers at Oregon State University (Nagy and Ullman, 1992: Chen et al., 1990) used the decision network to index the changing state of the evolving artifacts, so that the sequence, composition and dependence between decisions are described. It further describes the constraint development and propagation and the dependence on the design specifications. Information on the product is represented using two basic structures, the features of objects i.e., "object-attribute-value", and features of relations between objects, that is, "object1-object2-relationship-attribute-value", (Ullman et al., 1994). Here an "object" is defined as an assembly, a component, a feature, a human or other identifiable physical thing that is used to describe some physical aspect of the product being designed. Around the same time, Ganeshan et al. (1994) proposed a framework to capture design history of a design process,

[4]Decision Representation Language.

taking an example in the domain of spatial layout of small buildings. They represented the product being designed with an "objective-variable-value" structure over the design space. In their approach, the 'objective' is used to define design problems and to represent intermediate stages leading to the final solution; the decision-making process is an iteration of "focus-refine-evaluate-select-resolve" with an explicit linkage to each state (objective, variable, alternative) of the product. The design reasons are represented implicitly in the iteration.

Both of the above efforts have resulted in their significant progress in introducing the formal product data into the IBIS-based representation of design processes. However, as pointed out earlier, design has a large state space, where each state corresponds to a possible solution, either intermediate or final. The design process can be seen as a navigation from an initial state, the specification of the problem, to a final state, the proposed solution. Neither the "objective-variable-value" nor the "object-attribute-value" structure is sufficient for representing all product data in the whole state space explicitly. This results in the mappings between product data and process information not explicit and even a design process that is ephemeral and difficult to manage. Therefore, we attempt to capture the design history by introducing the concept of the *product data model* to represent the product data. Few related work records the histories of information on design teams. By contrast, we introduce the concept *agent* and *task* from the research (Jin and Levitt, 1993) on distributed artificial intelligence and organization theory to represent the *negotiation* activities in the collaborative design process.

3. Product Data Model

Product modeling technologies attempt to generate an information reservoir of complete product data to support various activities at different product development phases (Krause et al., 1993). The term product model can be interpreted as the logical accumulation of all relevant information concerning a given product during the product life cycle. Although a clear trend toward a wider usage of product models and a strong emphasis on product modeling processes is observed, no definite and commonly agreed product modeling approaches exist to date. To be effective and efficient, we focus on the modeling of product information on a smaller scale, namely, we address only a formal description of the issues, alternatives, and assignments in the design decisions concerning a product.

First, let us distinguish two basic concepts: *product data* and *product data model*. The term *product data* in this paper refers to the facts, concepts, or instructions about a product or set of products in a formal manner suitable for communication, interpretation, or processing by a human being or by automatic means (ISO, 1993a); *product data model* is an information model which provides an abstract description of the product data. The evolution of product data usually begins with an ill-defined need for a product and ends with exact specifications for production, use and retirement or recycling. Designers need to make decisions about product data at different levels of abstraction. The consequences of the presence of different levels of abstraction on product data must be explored. The product data are evaluated and changed through the different stages of a design process. The forward and feedback links among the various stages of the design process imply certain mappings and feedback among various parts of product data. Constraints play a crucial role in this regard. Product data must keep all constraints and differentiate between constructed versus derived geometric and non-geometric features. In addition, several design activities may have concurrence of accesses to product data. It is necessary that all activities performed during different phases of a process chain have identical data available to them concerning a particular subject. In this sense, this paper represents the design process on the basis of two assumptions, that is, each product datum, once it is created by any a design activity, is the same to all the design activities in which it is present; the constraints among various design activities imply the constraints among various elements of the product data.

The above requirements on product data for representing design processes and recording design histories lead to challenges in the development of the product data model in this paper. The most useful basis for developing the product data model is the resource constructs from STEP (ISO, 1993a), an international standard for the computer-interpretable representation and exchange of product data. Its objective is to provide a neutral mechanism capable of describing product data throughout the life cycle of a product, independent of any particular system. The nature of this description makes it suitable not only for neutral file exchange, but also as a basis for implementing and sharing product databases, and archiving. To specify and develop the STEP information models, a variety of tools for information modeling have been used. One of the tools, the data specification language EXPRESS (ISO, 1993b), focuses on the definition of entities, which are the *objects* of interest. The definition of an entity is in terms of its properties (attributes), which are characterized by specification of a domain and the constraints on that domain. The term resource construct refers to the collection of EXPRESS language entities, types, functions, rules and references that together define a valid description of product data. In this paper, all the text descriptions on data modeling are based on the terminology of EXPRESS.

As shown in Figure 1, the developed product data model includes five partial models. The requirement model supports the specifications of a product derived from an analysis of customer needs for the product. The entity *product-concept* from STEP part 44 is used in this model. A *product-concept* is a set of product features identified by the customers or derived from customers' needs. The *product-concept* is customer-oriented, while the *product* is engineering-design or manufacturing-design oriented. A product concept is essentially a marketing idea and includes customer-driven inputs. Therefore, the market-context is also defined as an attribute of the entity *product-concept*.

The function model defines the function structure of a product. No entity from STEP can be interpreted into this model. The function specifications are expressed in plain text in the description attribute of the *function* entity. The entity *function-relationship* defines the hierarchy between *functions*, while the entity *alternative-function-relationship* defines the relationship between base function and its alternatives. More detailed discussion about function analysis can be found in another paper (Taura, 1995) by one of the authors. In the solution principle model, the entities such as *physical-law*, *physical-phenomenon*, *physical-quantity*, and *working-domain,* have been formalized to represent physical effects, for example, the friction effect described by Coulomb's law ($F_F = \mu F_N$). The entity *physical-principle* defines the relationship between a *function* and the corresponding *physical-law* selected to fulfill the *function*, e.g., the friction effect used to fulfill the function 'transfer torque'. The entity *solution-principle* can then be specified by associated *part*s selected to fulfill the function and the parts' key features selected to define the *working-domain* of the corresponding *physical-law*. The entity *alternative-principle-relationship* is formalized to describe the alternatives of a *solution-principle*. However, the relationships between *solution-principle*s are not defined explicitly, since the *function-relationship* and/or *product-structure* imply them.

The development of the product structure model is based on the resource constructs from STEP part 44: 'product structure configuration' and STEP part 41: 'product description and support.' The product structure defines the different methods by which a product can be represented, as being made up of constituents. Product structure relationships are established among the assemblies and constituents that make up a product. The product structure (i.e., composition relationships) may be modeled mathematically by nodes representing assembly products and by directed links representing the "composed-of" relationship. Usually, two major data structures are used to represent product structure: bill-of-material and parts list. The bill-of-material structure is a structural description of a product in terms of its nested constituents, while the parts list structure is a structural description of a product in terms of a hierarchy of all distinct usages of its constituents. The product structure model in this paper supports both. The entities *product-definition*, *assembly-component-usage* and *alternative-product-relationship* are selected from STEP part 41 or part 44.

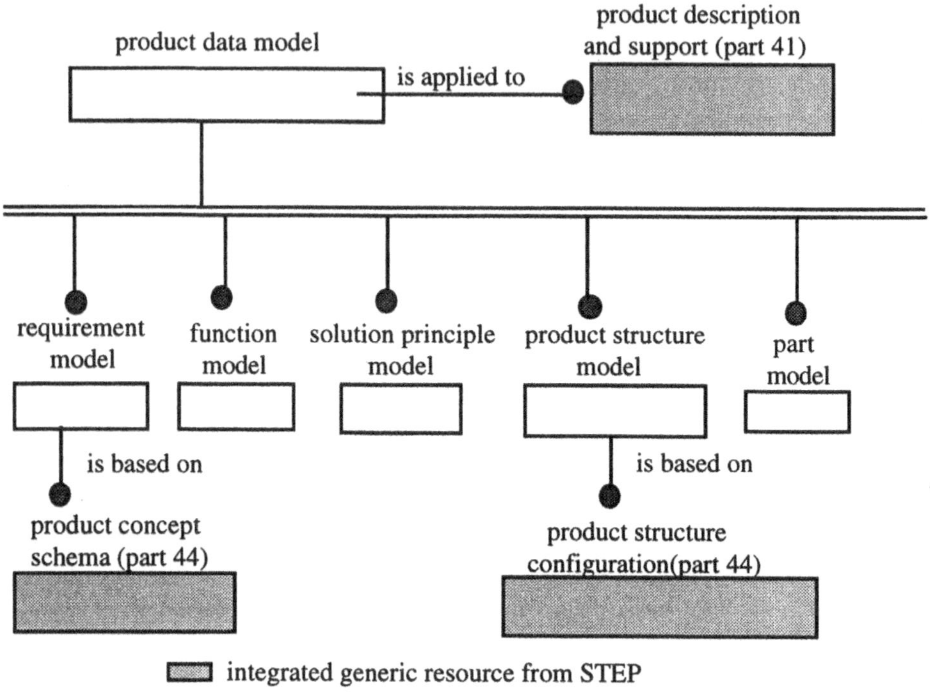

Figure 1: An overview of the product data model.

Figure 2 shows an IDEF1x graphic (NIST, 1992) of the main components for modeling the product data of a part. The resource constructs from STEP concerning the representation of material, tolerances, geometric model, and form features are interpreted into this model. This model supports the detailed design phase of a part. For an exact description of all the individual modules except the "parametric shape model," the readers are referred to the literature (Lei, 1994) written by one of the authors.

Contemporary shape representations are either geometric by CSG/B-reps or incompletely parametric by implicit form feature representations. A parametric shape model is desirable for the definition of all shape aspects into which the shape is divided, their configurations, and the constraints among them. Therefore, a generalized topology schema for parametric shape modeling has been developed. The proposed approach addresses the shape definition only with finite primitives and in a completely parametric manner. Introducing a new topological entity standing for the relationships among entities at different lower topological levels makes available a concise schema for the parametrization of both the shape aspects sharable for design and manufacture features and their dimensional relations rather than only dependency relations. It also ensures the universality of the schema for the parametrization of all shapes with or without free surfaces and makes the evaluation of such a parametric shape model into a B-rep model easy and unique. We have submitted a paper (Lei and Taura, 1995) about this parametric shape modeling to *CAD Journal*.

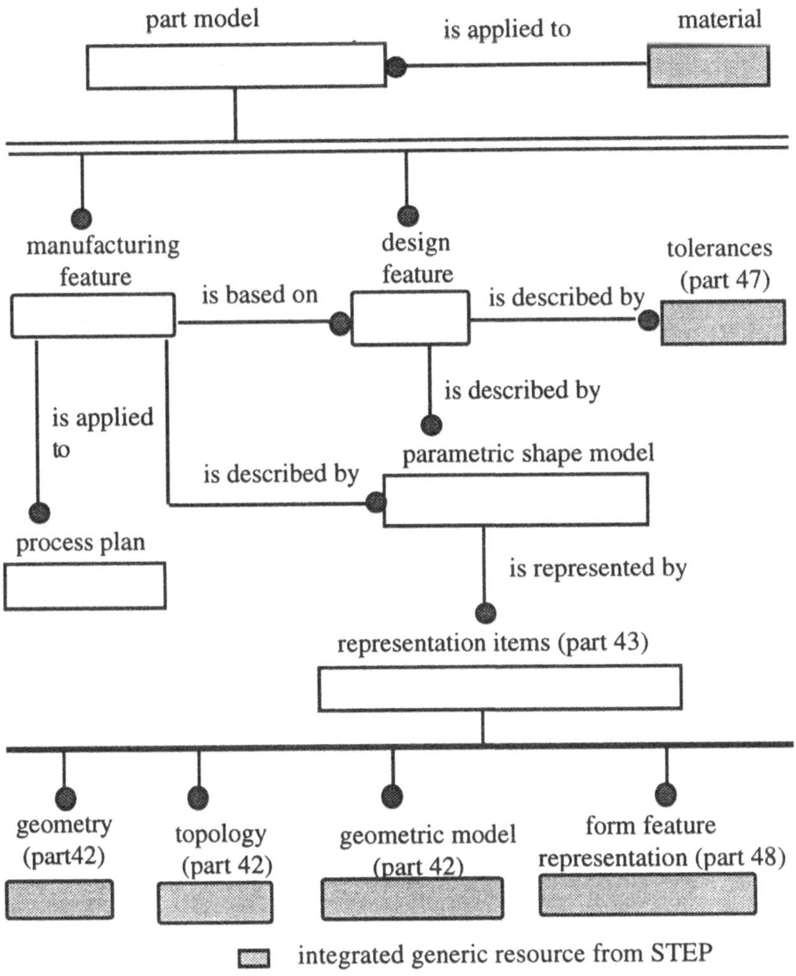

Figure 2. An IDEF1x graphic of the main components in the part model.

3. Formal Process Ingredients

To abstract and formalize the basic process ingredients, a collaborative design is generally regarded as an opportunistic process with a network of tasks. Each task records, in a temporal order, all the decision-making and conflict-resolution activities that fulfill itself. For the design deliberations of one decision, a process ingredient *activity* will be abstracted to accumulate the assertion, alternatives and assumptions of the decisions made on product data, the agent that made them, and the rationale used to make them. The mappings between process ingredients and product data take place only through three channels, that is, generations of the objectives of tasks via the ingredient *task*, determinations of the attribute values of a task's objectives via the ingredient *assignment*, and detection of conflict sources and constraints among tasks or among activities via the ingredient *negotiation*. All three channels use the product data model directly. Let us make an exact distinction between these ingredients.

3.1. ASSIGNMENT

Assignments are decisions on product data. As mentioned earlier, the collaborative design process is viewed here as the evolution of product data, which is the result of a series of decisions. Each decision is one step in the transformation of the product. In an assignment, the value of a single attribute, or values of coherent pieces of attributes if necessary, of an *entity* in the *product-data-model* is determined and recorded as assigned-data. These assigned data may be used as a proposed solution, an alternatives, an assumption, or a preference. In this sense, the ingredient *assignment* is the major channel through which process information interrelates and interacts with product data. The rationale behind these value determinations can be expressed formally by the entity *physical-law* defined in the above section, or PROCEDUREs specified in EXPRESS (ISO, 1993b) or informally by STRINGs, i.e., plain text. The rationale illustrates how and why the values are derived or determined. Once *assignments* are committed, all assigned data will be recorded as product data that can be shared by other activities and by other agents. The EXPRESS specifications of the entity *assignment* are as follows.

```
ENTITY  assignment;
        assigned-data          : product-data-model;
        rationale              : SET [1:?] OF rational;
        created-by             : (INV) activity;
        maker                  : agent;
        status                 : role;
END_ENTITY;
```

```
ENTITY   product-data-model;
         SUPERTYPE OF (requirement-model, function-model,
                       solution-principle, product-structure,
                       part-model):
END_ENTITY;

TYPE  rational = SELECT (physical-law, PROCEDURE, STRING) ;
END_TYPE;

TYPE  role = ENUMERATION OF (proposed, ready-for-alternative,
                             assumed, expected) ;
END_TYPE;
```

3.2. ACTIVITY

Activities are efforts that take place during a process. An *activity* aggregates the information generated by a single action during a process, including its objective, the decision, alternatives, and argument. There are different levels of granularity to aggregate the action information. On the finest level, information on micro decisions made at the rate of about one per minute is tracked. This shows that the fineness is necessary for completeness of information capture but is very difficult to implement and is even unrealistic for an operational system (Ullman, 1994). On the coarser project/program level of granularity, projects are defined as design activities performed by single-discipline teams whereas programs require team members from diverse disciplines. The design effort is normally seen in corporate product development plans. This is the level of information often handled by commercial systems such as IBM's Product Manager™/6000 and SDRC's DMCS[5]. Such systems may miss much important information concerning the alternatives considered, evaluations completed, and assumptions made. In this paper, action information is aggregated as activity at the attribute level of granularity. That means an activity is defined here as the design effort of a single agent on a single attribute, or coherent pieces of attributes of an *entity* in the *product-data-model*. Concretely, the *EXPRESS* specification of the entity *activity* is given below.

```
TYPE  criteria = ENUMERATION OF (cost, time, behavior, trade-off) ;
END_TYPE;

TYPE  state = ENUMERATION OF (admissible, optimal, rejected,
                              committed, retracted) ;
END_TYPE;
```

[5]Data Management and Control System.

```
ENTITY  activity;
        identification     : INTEGER;
        name               : STRING;
        goal               : task;
        assertion          : assignment;
        alternative        : OPTIONAL SET [1:?] OF assignment;
        argument           : criteria;
        maker              : agent;
        opportunity        : OPTIONAL SET [1:?] OF assignment;
        assumption         : OPTIONAL SET [1:?] OF assignment;
        status             : state;
END_ENTITY;
```

In the above specification, the attribute goal refers to the task whose objective is to be designed or determined in current activity. The task's objective will be defined in the next sub-section using any subentity of the *product-data-model*. The assertion slot records the proposed *assignment* on focused attributes of the current task's objective. The alternative slot corresponds to all the possible solutions to replace the assignment stored in the assertion slot. The attribute argument is described by the *criteria*, which explains why the *assignment* in the assertion slot is selected from all alternatives. The opportunity slot refers to the preference decision and the associated conflicts that make it unavailable. The attribute assumption assigns virtually all product data needed by executing current activity but which do not exist. The status of an *activity* can be one of the following: admissible, optimal, rejected, committed, or retracted.

3.3. TASK

A task is a set of pre-determined actions such as activities, negotiations, subtasks, and assignment-committing. Obviously, a collaborative design is not carried out by one designer in one session as a single task. In other words, collaborative design activities may occur in parallel. In this sense, the collaborative design scenario can be described by a network of tasks. *Activities* do not stand alone but are grouped into tasks. The task is determined both by the temporal ordering of *activities* and the interaction between grouped *activities*. The entity *task* in EXPRESS is specified below.

```
ENTITY  task;
        identification          : INTEGER;
        name                    : STRING;
        super-task              : OPTIONAL SET [1:?] OF task;
        objective               : product-data-model;
        working-activity        : LIST [1:?] OF  activity;
        charged-by              : agent;
        conflict-resolution     : OPTIONAL SET [1:?] OF negotiation;
        subtask                 : OPTIONAL SET [1:?] OF task;
        committed-activity      : SET [1:?] OF  activity;
END_ENTITY;
```

It has been pointed out in the discussion of the level of granularity that the objective of a task is to instantiate an entity in the *product-data-model*. Hence, here is the second channel through which a process ingredient interrelates with the product data model. The attributes of the instantiated object will be determined by a list of activities which are recorded in the working-activity slot as LIST instead of SET, since the temporal order of the activities is also useful information for the design history. A task consists of at least one working activity. If multiple working activities exist in a task, the conflict-resolution slot can optionally store all possible actions to resolve the possible conflicts among the working activities using the entity *negotiation* formalized in the next section. After conflict resolutions, the resulting new set of consistent activities can be recorded in the committed-activity slot. All assigned data created in each *assignment* of the committed activities, either as assertion or as alternatives, can be committed into product data for further common sharing by the responsible agent of the current task, who is appointed in the charged-by slot. Only this agent has access to commit *assignments* into product data in the task level, although multiple agents can submit different *assignments* in the activity level to fulfill the objective of the current task.

3.4. NEGOTIATION

Negotiations here refer to the actions initiated to resolve conflicting activities. There are three kinds of conflicts. The simplest one is the serial collaboration between activities, that is, one activity can take place only after the decision of another activity is ascertained. In such a case, if both activities are within the same task, they can be arranged in a corresponding serial manner by the responsible agent. Otherwise, either assumptions or negotiations can be raised by the suspended activity. The second one is the conflict assignments for the same attribute from different perspectives, i.e., a single attribute of the objective of a task may be determined by multiple agents in different disciplines in multiple distinct *activities*. The last and most complex one is the dependent assignments, viz., some constraints representing the relationship among the attributes of different task objectives are invalid according to the current assignments by all corresponding activities. Once the identity of the conflicting perspectives is known, the negotiation process must be initiated to resolve conflicts.

The product data model plays a key role in such situations. As pointed out earlier, we represent the design process on the basis of two assumptions, that is, each product datum, once created by any design activity, is the same to all design activities in which it is present; and the constraints among various design activities imply the same constraints among various *objects* of the product data. If the data assigned by an activity are committed into a shared product database supported by the product data model described above, the two assumptions are admissible. The product data model written in EXPRESS is certainly object-oriented and EXPRESS supports the definition of constraints among the attributes of different entities by RULEs and PROCEDUREs. Therefore, the first assumption, together with the uniqueness of an *object,* assure the detection of the first and second kinds of conflicts. Usually, the dependence among the attributes in the same entity is defined by defining 'DERIVE' attributes in an entity of the *product-data-model*, while the attribute dependencies among multiple entities are defined as associated constraints in the form of RULEs including PROCEDUREs. This assures the detection of the third kind of conflict. In a word, conflict detection and constraints propagation can be controlled through product data, because conflict sources and/or constraints can be formally and explicitly expressed by the *product-data -model* .

```
ENTITY  negotiation;
        identification          : INTEGER;
        name                    : STRING;
        conflict-source         : SET [1:?] OF product-data-model;
        conflict-activity       : SET [2:?] OF activity;
        conflict-resolution     : SET [1:?] OF activity;
        negotiator              : SET [2:?] OF agent;
        note                    : OPTIONAL STRING;
END_ENTITY;
```

As shown in the above specification, at least two *agents* recorded in the negotiator slot take part in negotiation. In addition, at least two conflicting activities are involved to be resolved in a negotiation and can be stored in the conflict-activity slot. The conflict-source slot records either invalid constraints or conflicting attributes of an *object* defined by the *product-data-model.* The result of a negotiation for conflicting activities may be one or a set of new *assignments* accepted by all of the negotiators and can be represented explicitly by a set of *activities* in the conflict-resolution slot, since the new *assignments* in these situations also involves the same properties of an *activity* such as alternatives, and assumptions. The note slot is ready as an optional property to record some other information that must be recorded but cannot be formally represented in the above-mentioned slots, because negotiations may consist of some very complex deliberations.

3.5. AGENT

An agent here refers to a combination of human and software information storage and processing. The combination ranges from an agent that is human with a software interface to interact with other agents, or an agent completely implemented in software. The agent description includes role characteristics such as position in the team hierarchy; authority for design, approval, and coordination tasks. A concrete description of the ingredient *agent* is shown in the following EXPRESS specifications.

```
ENTITY  agent;
        identification          : INTEGER;
        name                    : STRING;
        role                    : organization-role;
        in-charge-of            : OPTIONAL SET [1:?] OF task;
        proposed-activity       : OPTIONAL SET [1:?] OF activity;
END_ENTITY;
```

Here, the role characteristics are expressed by the entity *organization-role*, which is selected from the management_resources_schema in STEP part 41. The other two attributes define the related tasks and activities charged by the current agent.

4. Example

The above-mentioned product data model is developed in EXPRESS as a STEP application protocol. It includes the detailed specifications of schemata of the five partial models discussed above. The normative references for these schemata are as follows: STEP part 41, 42, 44, 45, 47 and 48. The data model is implemented on an object-oriented database called ONTOS in C++. All product data concerning an example shaft are instantiated on the corresponding database schema (Lei, 1994).

Based on the above-mentioned formal process ingredients, a design history base, as an extension of the product database, is under development to record the design history of Sony's color video printer UP-5000. Two of the authors have analyzed in detail the Sony's current product planning and design process, taking the development of this printer as an example (Numata and Taura, 1995). The first assembly selected to be implemented is the paper-handling system of the printer. We have focused on one particular kind of paper transport, namely, that which uses pinch rolls to isolate one sheet from the other sheets piled at the paper entrance. Figure 3 shows the example instances of the process ingredients.

The design of the isolation mechanism in the paper-handing system involves six tasks: the design of pinch roller, belt, rubber roller, gate roller, press plate and floor guide. Within the *task* for pinch roller design, task1 (#4 in Figure 3), three independent activities are planning for the inner-diameter, for the outer-diameter, and for the width. Its objective is to instantiate the entity *pinch-roller* in product data model with an empty instance, pinch-roller1 (#10 in Figure 3). Within the *activity* for determining the width of the roller (#3), two assignments, assignment1 and assignment2, are possible.

```
#3 = activity1
  name: "set roller width";
  goal: #4;
  assertion: #1;
  alternatives: (#2);
  criteria: cost;
  opportunity: ();
  assumption: ();
  status: admissible;
  maker: #8; //Taroh;
```

product data model

```
ENTITY pinch-roller;
  d : REAL;
  d1: REAL;
  b : REAL;
END_ENTITY;
```

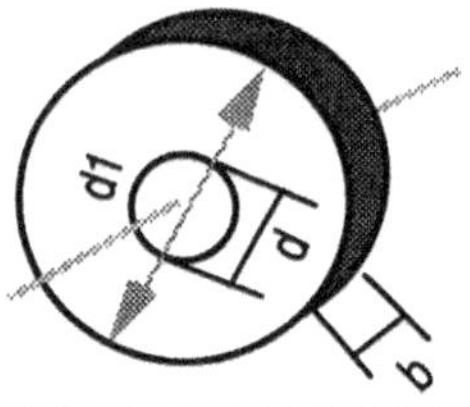


```
#10 = pinch-roller1
  d: (NULL);
  d1: (NULL);
  b: (NULL);

#1 = assignment1
  assign-data : (#10:b)=200
  rationale: "catalog"
  maker: Taroh

#4= task1
  name: "design pinch-roller";
  super-task: #1000; // paper handling mechanism
  objective: #10 ;
  working-activity: (#3, #300, #301)
  conflict-resolution: #5;
  subtask: ();
  charged-by: #1100; // agent3=Tom
  committed-activity: (#7, #300, #301)

#2 = assignment2
  assign-data : (#10:b)=240
  rationale: "experience"
  maker: Taroh

#6 = assignment3
  assign-data : (#10:b) = 160;
  rationale: (#100: W )+11;
            // belt width
  maker: #9; //Hanako;

#5=negotiation1
  name: conflicts about pinch-roller's width;
  conflict-source: #10
  conflict-activity: (# 3, #600);
      // #600's assertion is #6
  conflict-resolution: (#700);
      // assertion is #7
  negotiator: (#1100, #8, #9)

#7 = assignment4
  assign-data : (#10:b)=180;
  rationale: "trade-off"
  maker: (#8, #9); // Taroh, Hanako

#8 = agent1
  name : Taroh;
  role: engineer
  in-charge-of: (#4); //design pinch-roller
  proposed-activity: (#3);

#9 = agent2
  name : Hanako;
  role: engineer
  in-charge-of: (#1000);
         //design belt
  proposed-activity: (#600);
```

Figure 3. Example instances of the process ingredients.

The agent responsible for roller design, Taroh, asserts assignment1 in consideration of the cost criteria. However, the agent in charge of belt design, Hanako, suggests another assignment3 (#6), because of the constraint between the roller width and belt width. Then negotiation1 (#5) occurs between Taroh and Hanako and assignment4 (#7) is decided as a "trade-off". The product data model in Figure 3 is provided only for easier understanding of the design process entities. The real product data model in the database is more abstract.

5. Conclusions

A product model-oriented approach for formal representation of the collaborative design processes has been proposed to capture design histories. One of the important features of this proposed approach is introduction of the product data model to control the explicit mappings between process ingredients and product data. To be more precise, the product data model supports the generation, exploration and navigation of the objective in each task/activity in a formal manner; the product data model also explicitly represents the evolution and alternatives of product data from objectives focused through all intermediate assignments to final specifications at any state of its recorded history, beginning with the general definition of customer's requirements for a product and ending with the exact specifications for production, use, and retirement or recycling; in particular, constraints among product data defined by the product data model imply constraints among tasks , and the product data model facilitates the detection of conflict sources and propagation of the constraints among activities.

In addition, the proposed representation regards a collaborative design as an opportunistic process with a network of tasks. In each task the temporal order of the decision-making and conflict-resolution activities has been recorded. For the design deliberations on one decision, the process ingredient *activity* has been specified to record the assertion, alternatives and assumptions of the decisions made on product data, the agent that made them, and the rationale used to make them.

However, all the data models are assumed to be statically defined. There are no STEP mechanisms for modifying a product model while it is in use. Further research will focus on dynamic definition of the data model to reflect the dynamic nature of design. Another future work will be a more formal representation of the design rationale in the framework of the proposed approach by means of a formal description of the agent's knowledge and communication patterns. A concurrent design environment will be developed taking the developed data model and database as the information infrastructure.

Acknowledgments

The work reported in this paper is supported partly by Japan Society for Promotion of Science. Any opinions, findings, conclusions and recommendations expressed are those of the authors, and do not necessarily reflect the views of the sponsoring agencies.

References

Banares-Alcantara, R.: 1991, Representing the engineering design process: two hypotheses, *Computer-Aided Design*, **23** (9), 595-603.

Brown, D. C.: 1994, Rationale in design, *in* P. W. H Chung and R. Banares-Alcantara (eds), *AID'94Workshop on Representing and Using Design Rationale*, pp. 1-3.

Chen, A., McGinnis, B. and Ullman, D. G.: 1990, Design history knowledge representation and its basic computer implementation, *Proceedings of the Second International ASME Conference on Design Theory and Methodology, ASME DE,* 27, pp. 157-184.

Chung, P. W. H. and Goodwin, R.: 1994, Representing design history, *in* J. S. Gero, and F. Sudweeks (eds), *Artificial Intelligence in Design '94,* Kluwer, Dordrecht, pp. 735-752.

Conklin, J. and Begeman M. L.: 1988, gIBIS: a hypertext tool for exploratory policy discussion. *ACM Transactions on Office Information Systems,* **6**(4), 303-331.

Ganeshan, R., Garrett, J. and Finger, S.: 1994, A frramework for representing design intent, *Design Studies,* **15**(1), 59-84.

ISO: 1993a, *Standard for the Exchange of Product Model Data, ISO DIS 10303*, International Standards Organization.

ISO: 1993b, *Description Methods: the EXPRESS Language Reference Manual, ISO DIS 10303 Part 11* International Standards Organization.

Jin, Y. and Levitt R.: 1993, i-AGENTS: modeling organizational problem solving in multi-agent teams, *Intelligent System in Accounting, Finance and Management,* **2**, 247-270.

Klein, M.: 1993, Capture design rationale in concurrent engineering, *IEEE Computer*, **1**, 39-47.

Krause, F-L., Kimura, F., Kjellberg, T. and Lu, S.: 1993, Product modeling, *Annals of the CIRP,* **42.**

Lee, J.: 1991, Extending the Potts and Burn model for recording design rationale, *Proceedings of the 13th International Conference on Software Engineering.*

Lei, B.: 1994, A data model for the integration of CAD/CAPP/NC, *Technical Report*, Laboratory for Machine Tools and Production Engineering (WZL), Aachen University of Technology, Germany.

Lei, B. and Taura, T.: 1995, Parametric shape modeling of mechanical parts: A generalized topology approach, *CAD* (submitted).

Mostow, J.: 1985, Toward better models of the design process, *AI Magazine,* **6**(1), 44-57.

Nagy, G. L. and Ullman D. G.: 1992, A data representation for collaborative mechanical design, *Research in Engineering Design,* **3**(4), 595-603.

NIST: 1992, *DEF1X (ICAM Definition Language 1 Extended) Integration Definition for Information modeling*, FIPS PUB XXX, NIST.

Numata, J. and Taura, T.: 1995, A network system for knowledge amplification in the product development process, *IEEE Engineering Management* (submitted).

Potts, C. and Brun, G.: 1988, Recording the reasons for design decisions, *Proceedings of the 10th International Conference on Software Engineering,* CS Press, pp. 418-427.

Rittel, H. W. J. and Webber M. M. J.: 1973, Dilemmas in a general theory of planning, *Policy Sciences,* **4**, 155-169.

Rosenman, M. A., Gero, J. S. and Maher, M. L.: 1994, Design intent and multiple abstraction, Knowledge-Based Research at the Key Centre of Design Computing,

University of Sydney, *Working Paper*, Key Centre of Design Computing, University of Sydney, pp. 20-24.

Smithers, T., Conkie, A., Doheny, J., Logan, B., Mollington, K. and Tang, M. X.: 1989, Design as intelligent behavior: An AI in design research programme, *Research Paper DAI 426*, University of Edinburgh, UK.

Taura, T.: 1995, Design science for functional design process modeling, *Proceedings of the International Conference on Engineering Design, ICED'95* .

Thompson, J. and Lu, S.: 1990, Design evolution management: a methodology for representing and using design rationale, *Proceedings of Second International ASME Conference on Design Theory and Methodology*

Ullman, D. G.: 1994, Issues critical to the development of design history, design rationale and design intent systems, *in* T. K. Night and F. Mistree (eds), *Proceedings of the Sixth International ASME Conference on Design Theory and Methodology*, pp. 249-258.

Ullman, D. C., Herling D. and Sinton, A.: 1994, Analysis of protocol data to identify product information evolution and decision making process, *Analyzing Design Activity Delft Protocol Workshop*, pp. 67-82.

16

DESIGN PROCESS METHODS: DISCUSSION

Research Paradigms

LEO JOSKOWICZ

IBM T. J. Watson Research Center, USA

1. Introduction

Engineering design is a complex, evolving activity, involving many kinds of processes aimed at turning requirements and ideas into products. Concept development, concept evaluation, system-level integration, detailed design, prototyping, manufacturing, production, and marketing, are all part of an increasingly demanding iterative design cycle. Global competition requires shorter and more efficient design cycles, more systematic evaluation of possible solutions, and a better understanding of the relations between the many factors affecting the final product. Faced with these demands, research in computer-aided design has rapidly expanded its scope from traditional drafting programs to systems capable of supporting many of the diverse activities associated with product design.

Understanding and characterizing the different aspects of the design process is the key to the development of a new generation of computer-based design tools. While design methodology has been the subject of study for more than half a century, only recently have formal design process methods emerged from research in CAD. As is the case in other fields, such as finance, communications, and manufacturing, the focus on computational methods has not only provided a better and more systematic foundation to the field, but it has contributed to its evolution, and is likely to give rise to new paradigms.

The goal of this chapter is to briefly reflect on the state of the art of recent research in formal design process methods, identify trends and paradigms, and discuss key technical issues. Most of the background material for this paper is drawn from presentations and discussions at this workshop, although it reflects only the author's opinions and views.

2. Research Methodology

Most researchers and practitioners view design as a series of interrelated processes that must be studied and characterized to develop adequate CAD tools to support or automate significant aspects of the design activity. These design processes cover a wide spectrum, ranging from the well-defined and highly structured to the poorly understood and difficult to isolate. These processes and the interactions between them are highly dynamic, drawing on different sources of knowledge at different levels of abstraction.

The three major components that have been identified are the design knowledge, the domain knowledge, and the derivation process. The design knowledge is generic, domain-independent, and includes techniques for identifying customer needs, methods for product concept generation, and mathematical optimization, to name a few. The domain knowledge is specific to the type of product being developed and its application domain; it includes part catalogs, physical equations of behavior, and previous designs. The derivation process includes the methods by which new or improved designs are produced, such as transformation and refinement operators, deduction, and induction.

There is a broad consensus about the importance of each of these components. Design methods for the better understood aspects of the design process, such as part catalog selection, collaborative design information exchange, and multi-objective optimization, have been developed and are constantly improving. The current research emphasizes collaborative design methods, conceptual design support, and design rationale capture, among others. Knowledge representation issues related to the context of the design and the explicit representation of form, function, and behavior have also been recently investigated.

A major area of disagreement, addressed explicitly or implicitly in the papers, relates to the nature of the design process itself and the best way to formally represent it. Some researchers propose to use formal logic and postulate induction, deduction, and abduction as the basic design derivation and modification mechanisms. Others take a more empirical stance and attempt to follow established design methodologies as much as possible. Some propose an evolutionary approach and try to reproduce as closely as possible human decision-making. Depending on the approach, deriving properties of the design process can be seen as deriving pragmatic, rule-of-thumb observations, or proving properties of formal systems. These considerations yield very different formalizations.

A related issue is the degree of specificity of the theory with respect to the design domain and process. The question is to determine what is common and what is distinct in architectural design, mechanical design, and VLSI design, to name a few examples. In other words, in what sense is the design of a house similar or different to the design of a car or a household electric appliance? Some fields and design activities are more generative, some emphasize enumeration and selection, while others rely heavily on adaptation of previous designs. Some must follow a highly structured set of rules, while others rely on unquantifiable judgements, such as aesthetics. Yet another difference is whether the design is product-driven (products designed to meet specific consumer needs), technology-driven (products designed to push a new technology), or process-driven (products whose characteristics rely heavily in a production process). These differences are not merely a question of the degree of creative, innovative, detailed, or routine design involved: they are at the base of the discipline and the design process itself.

These issues have a major impact on how researchers develop formal and computational theories of the design process. The definition, systematization, and exploration of the design space, its focus and representation, and the derivation processes heavily depend on the genericity and domain-independent assumptions that are made. This helps to explain why some works emphasize mathematical formalization (logic, optimization, etc.), while others take a procedural or behavioral approach, focusing on reproducing the input/output behavior of the human design process. Approaches that attempt to mimic human designers are based on empirical studies of the design process through user interviews and design protocols.

This diversity in assumptions, emphasis, and focus makes comparing and evaluating the proposed methods very difficult. Some researchers have proposed to use benchmarks as an objective means to compare different approaches. Benchmark studies are indeed useful to understand the scope, coverage, and efficiency of a set of methods with respect to a task. However, they are only suited to relatively mature fields, with well-defined and well-understood problems, such as comparing the performance of two computers. For design process methods, we cannot even begin to agree on what the problem is, let alone define what criteria should be used to compare them.

3. Design Process Paradigms

In reviewing current work on design process methods, it is interesting to note that most of the proposed methods are build around a single paradigm, with the assumptions, simplifications, and specific focus (not always explicitly stated) that this entails. I will briefly review the two papers presented on design process methods. The goal is to identify the paradigm upon which they are based and make their assumptions explicit.

Brazier, Van Langen, and Treur present a logical theory of design based on formal semantics for both the static (knowledge) and dynamic (reasoning) aspects of design. The theory, which is based on many-sorted first order predicate logic, partial temporal models, and non-monotonic reasoning, is generic and domain independent. The authors justify their approach with the paradigmatic observation:

> "Design tasks typically reason with incomplete and inconsistent knowledge of requirements and design object descriptions; they reason non-monotonically

with and about, for example, (default) assumptions, contradictory information, and new design knowledge."

Interestingly, the authors directly go on to describe the technical details of the formal semantics, which they claim provides a means to model design strategies, without explaining why such a formal semantics is desirable.

Lei, Taura, and Numata present a product model-oriented approach, where the data associated with the product at the different stages of the product life cycle serves as the key organizing principle. The model supports the generation, exploration and navigation of the design objective in each task or activity, and the evolution and alternatives in the design history. This paradigm is summarized by the following observation:

"The collaborative design process can be viewed essentially as the evolution of product data, which is the result of a series of decisions."

The central role of explicit modeling of the evolution alternatives and constraints of the product data justifies the emphasis on the product data model. The computer implementation, based on the language STEP (Standard for the Exchange of Product Model Data) targets electro-mechanical product design.

4. Conclusion

As the reader can appreciate, it is difficult to compare or even put these theories in perspective with respect to each other. In my opinion, there should be a greater effort to precisely characterize the scope and domain of application of each theory, and identify both their theoretical and pragmatic limitations, which are not always the same. For example, a theory can be very expressive but practically very difficult to compute with, or vice-versa. The critical assumptions must be stated explicitly, and greater effort should be put in clearly stating the paradigm upon which it is based, the domain where it is most likely to be useful, and the domain or type of design where it is not useful.

Following this methodology will allow us to begin assembling a "toolbox" of design methods, while the "ultimate" theory (if such a theory indeed exists) is being worked on. It will allow us to identify areas that require more research, and problems that deserve more attention. It will provide practitioners with an understanding of what is and is not presently possible, and useful approaches, if not programs, to address their problems.

Conversely, and on a more pragmatic level, we should also examine existing needs and identify the current bottlenecks in the design of a particular class of products and domains. This will help us identify what developments would yield the maximum benefit, and help motivate research both in the short and the long term.

Closing Discussion

ADVANCES IN FORMAL DESIGN METHODS FOR COMPUTER-AIDED DESIGN

JOHN GERO
University of Sydney, Australia

1. Introduction

The introduction to the closing chapter of the proceedings of the workshop on Formal Design Methods for Computer-Aided Design (Gero, 1994) commences with the following paragraphs, which are still applicable.

The primary axiom of formal design methods is that design is a process, ie, it is temporally based and various distinguishable activities can be ascribed as occurring during the process. The use of formal design methods does not imply that design, when carried out by humans, is based on these formal methods. Rather, the use of formal methods initially provides a framework for our notions about design. Just as the use of scientific methods to study emotion does not make emotion a scientific process, so the use of scientific methods to study design does not make design a scientific process. However, once we have formal design methods we can conceive of uses for them other than as simple descriptive devices.

What are the possible roles that formal methods can play in design and in computer-aided design in particular? Formal methods in other disciplines have been found to be useful in a variety of ways and, by analogy, we are able to enunciate roles for formal methods in design. An incomplete list of potential roles includes:

1 informing us about design as a process;
2. providing a framework for comparisons amongst alternate design processes; and
3. providing a basis for the development of design tools.

To this list we can add the following:

4. provide a basis for distinguishing different kinds of design processes.

2. Informing Us About Design As a Process

Design is considered to amongst the most complex and most intellectual of human activities. It is the basis for the change of the physical world we inhabit. As such it is surprising that it is neither well understood nor well characterised.

Design is widely regarded as a process during which the designer carries out distinguishable activities. These activities can be characterised through a formal structure which can be used to provide a framework to assist in the understanding of what design is. Introspection, retrospection and protocol (Gero and McNeill, 1996) studies provide a basis for this use of formal methods. A formal method is proposed as a structure capable of providing a framework for our ideas about design. That structure is often found to be satisfactory in providing a framework for some aspects only of our understanding of design. However, other aspects may not be accounted for, hence other structures are needed to augment or replace the existing ones.

Formal design methods provide the opportunity to characterise design as a process in a uniform manner using concepts and terminology which transcend the individual designer and gives primacy to the processes of designing. Formal design methods are often taught as part of either a design theory course or in a design course as a process of designing. There appears to be a divide between the technology-oriented design courses such as in engineering and those in the human-oriented design courses such as in industrial design and to a lesser extent architecture. The former courses are beginning to embrace formal design methods whilst the latter less so.

Early formal methods proposed the analysis-synthesis-evaluation model of design. Whilst today this is considered to be an inadequate model it has provided the framework for the development of many other process models. Other characterisations, such as that of function-behaviour-structure, of routine/non-routine design, and of case-based/compiled knowledge design have all added to our armory to aid our understanding of design. What these formal characterisations have shown is that design is a highly complex activity which needs many characterisations for its understanding.

3. Framework For Comparisons Amongst Alternate Design Processes

The field of design research is populated by researchers who come from very varied backgrounds. They come from the design disciplines themselves: architects, engineers, industrial designers, and so on. They also come from such disciplines as psychology, computer science, history and philosophy. Given this variegated background one role of formal methods is to provide a framework which allows comparisons to be made amongst alternate design processes or alternate descriptions of the same design processes. It is not intended here to provide a framework which allows for the comparison of alternate design processes. A few examples will suffice to illustrate the ideas involved. Whilst a model of design based on concepts from optimization appears to bear little relation to one based on constraint satisfaction, they can both be categorised under a formal approach based on search. The notion of search, itself, carries with it concepts associated with a space to be searched. This introduces the need to characterise design spaces and provides the opportunity to extend our understanding of what design spaces are and how they may be created and examined.

Whilst there are well-defined frameworks derived from both within and without design research, none appears to be able to provide both the breadth of abstractions and the depth of process structure needed to be all encompassing. The *function-behavior-structure* framework (Gero, 1990; Umeda et al., 1990) has proven to be both useful and durable as a means of articulating fundamental differences in the characterizations of design activity unrelated to individual processes which execute that activity.

The other common framework which is process-oriented is the *search–exploration* framework. Here search is considered to be that set of design processes which assumes that design can be treated as operating within a fixed space of possible designs and any design process can be characterized as searching for appropriate solutions. From a computational viewpoint this is a very attractive approach since it readily maps onto various standard algorithms such as numerical optimization in the numerical world and constraint-based reasoning in the symbolic world. On the other hand, exploration is that set of design processes which manipulate design spaces which are then subsequently searched.

Both these frameworks are suitable at an abstract level for providing a structure for alternate design processes but are not concrete enough to allow any form of detailed comparison. . The function-behavior-structure framework is useful in determining the locus of a process within a design activity since it articulates the six, common different design activities:

1. design formulation
2. design synthesis
3. design analysis
4. design evaluation
5. design reformulation
6. design documentation.

The primary advance in this area has been the increasing recognition that the search–exploration framework provides a means of distinguishing design from many other activities which it appears to look like; activities such as problem solving and planning. The effect of this has been to allow such design processes as case-based design and a variety of 'creative' design processes to fit into this framework in order to make comparisons between them at this level of abstraction. This comparison is not via benchmarking but in terms of locus in the design process and in terms of the capability of the process to produce the stated output.

4. Basis for the Development of Design Tools

There are two complementary views taken about formal design processes. The first is that formalizing such processes helps us understand design as a series of processes better. The second is that formal design processes are the basis for the development of design tools. A design tool is a computer program which takes a specified design task and provides one of three levels of design aid:

1. provides active support in the form of a complete solution to that design task, for example synthesising or selecting a particular component which satisfies a set of constraints;
2. provides active support in the form of the beginnings of a solution to that design task, for example a conjecture about a possible direction or path to follow; or
3. provides passive support for a design task, for example an analysis or evaluation related to that design task.

This workshop has demonstrated that a variety of novel processes can be introduced into design.

5. Distinguishing Different Kinds of Design Processes

For many computer-aided design simply refers to the process of documenting a design, thus for most practitioners today CAD stands for either computer-aided drafting or computer-aided analysis although there is an increasing realization that there is more. Formal approaches allow us to distinguish different kinds of design processes according to both their locus in an overall view of design and their methodology.

There appear to be three bases of design methods:

1. those based on empirical results;
2. those based on axioms; and
3. those based on conjectures.

This third category can be further subdivided into:

3.1 those where the conjecture is founded on an analogy with human design processes, and

3.2 those where the conjecture is founded on concepts other than human design processes.

Surprisingly, there are very few empirical results derived from human designers on which to base a design method. As a consequence there are very few formal design methods which are directly based on human design processes. Just as there very few design theories to explain the phenomena of human designing. The closest example of this class of design methods is that presented by Rodgers et al. in this volume although it is not strictly based on direct empirical evidence.

Formal design methods based on axioms are more common than those based on empirical results but even here there are remarkably few and the discussion by Rudolph in this volume examines some of the reasons why there are so few. Such axiom-based formal methods include those presented by Rudolph and by Brazier et al. in this volume.

By are the most common basis for current formal design methods are conjectures, conjectures of both kinds listed above. Design by combination is a good example of a formal design method based on a conjecture which is itself founded on an analogy with a human design process which is claimed to be of that kind. This class of design methods is the most common even though not all the conjectures are well-founded. Amongst this class is found the formal methods of: design by generate-and-test, design by refinement, design by analogy, case-based design, top-down methods, etc. Examples of this class of formal design methods in this volume include those presented by Kalay and Carrara, and Grabowski et al.

The second class of conjecture-based design methods include the mathematically-based methods and the large corpus of design systems which use a heuristically derived system architecture as the framework. Examples of this class of formal design methods in this volume include those presented by Maher et al., Gero and Kazakov, Brown and Cagan, Zozaya-Goristiza and Estrada, Lenart, and Lei et al.

6. Conclusions

Formalizing design methods allows them to be better understood before they are used. It provides the basis for any mode of comparison between methods whether these comparisons be based on the methods' complexity, ontology, epistemology, teleology or applicability. It becomes possible to examine the method separately from the results it produces and to determine its behavior. It becomes possible to consider a method's expressive power and to determine whether it is a strong but specialised method or a weak but general method.

Perhaps the most significant aspect of formalizing design methods is that it then becomes possible to implement them as computer programs which radically changes both their character, their testability and their applicability. Without such formalization computerization becomes extremely difficult if not impossible. Computer implementations also allow for methods to be tested empirically on a wide variety of design problems.

The eleven contributions found in this volume provide a guide to the state-of-the-art of formal design methods for computer-aided design. They demonstrate both the range and depth of such methods. They also demonstrate that formalizing design methods is still in its early stages and more needs to be done to derive all the benefits from such formalizing.

References

Gero, J. S.: 1990, Design prototypes: A knowledge representation schema for design, *AI Magazine* **11** ((4): 26–36.

Gero, J. S.; 1994, Formal design methods for computer-aided design, *in* J. S. Gero and E. Tyugu (eds), *Formal Design Methods for Computer-Aided Design,* North-Holland, Amsterdam, pp. 353-359.

Gero, J. S. and McNeill, T.: 1996, An approach to the analysis of design protocols, *Research in Engineering Design* (submitted).

Umeda, Y., Takeda, H., Tomiyama, T. and Yoshikawa, H.: 1990, Function, behavior and structure, *in* J. S. Gero (ed.), *Applications of Artificial Intelligence in Engineering V: Design,* Springer-Verlag, Berlin, pp. 177-193.

AUTHOR INDEX

Batch number: 10761599

Printed by Printforce, the Netherlands